MASTERING GPS FLYING

PHIL DIXON

SHERWOOD HARRIS

D0107456

McGraw-Hill

New York Chicago San Francisco Lisbon London Madrid Mexico City
Milan New Delhi San Juan Seoul Singapore Sydney Toronto

The McGraw·Hill *Companies*

Library of Congress Cataloging-in-Publication Data

Dixon, Phil, date.
    Mastering GPS flying / Phil Dixon, Sherwood Harris.
       p. cm.
    Include bibliographical references.
    ISBN 0-07-141669-2
    1. Instrument flying.   2. Global Positioning System.   I. Harris, Sherwood.   II. Title.

    TL711.B6D58 2004   2004054699

1 2 3 4 5 6 7 8 9 0   DOC/DOC   0 1 0 9 8 7 6 5 4

ISBN 0-07-141669-2

No copyright claim is made for any material that has been taken from U.S. government sources. The aeronautical charts and related data appearing this book are for illustration purposes only and are not for use in navigation.

The sponsoring editor for this book was Scott Grillo, the editing supervisor was David E. Fogarty, and the production supervisor was Sherri Souffrance. It was set in Bembo by Deirdre Sheean of McGraw-Hill Professional's Hightstown, N.J., composition unit. The art director for the cover was Anthony Landi.

Printed and bound by RR Donnelley.

This book was printed on recycled, acid-free paper containing a minimum of 50% recycled, de-inked fiber.

McGraw-Hill books are available at special quantity discounts to use as premiums and sales promotions, or for use in corporate training programs. For more information, please write to the Director of Special Sales, McGraw-Hill Professional, Two Penn Plaza, New York, NY 10121-2298. Or contact your local bookstore.

CONTENTS

1 Introduction 1

2 Mastering GPS Flying in Eight Lessons 3
    Flight Lessons 3
        Shortcuts That Don't Shortchange-1 5
    Preflight and Postflight Briefings 6
    Background Briefings 6
    Learning the Keystrokes 7
    Docking Stations 8
        Shortcuts That Don't Shortchange-2 9
    Black Box Practice..10
    Don't Overlook This Fundamental.............................10

3 All You Need to Know About GPS Technology.............11
    Don't "Read" the Manuals! 11
    How GPS Works..12
        Useful GPS Terms 14
    Great Circle Navigation...14
    VFR Use of GPS 15
    RAIM...16
        Shortcuts That Don't Shortchange-3 17
    How Accurate Is GPS?...17
    How WAAS Works 18
    LAAS Progress 19
    Approaches 20
    Lower Minima 20
        Black Box Practice-1 22

4 Buttons, Knobs, and Switches 23
    Forget About American Samoa! 23
    Power Up 25
    Do-It-Yourself Checklists 26
    Direct-To 27
    OBS Mode 28
    Entering Flight Plans 28
    Waypoint Autosequencing 29
    Waypoint Alerting and Turn Anticipation..............29
    Frequency Autoselection 30
    Nearest Function 30
    Vertical Navigation (VNAV) 31
    Oops! I Punched the Wrong Button.........................32
        Shortcuts That Don't Shortchange—4...............*32*
        Black Box Practice—2 ....*35*

5 How to Plan a GPS Flight...37
    GPS Planning with DUATS.....................................37
    Required VOR Backup for GPS 38
    Sample GPS Flight Plan 39
    Rehearsing the Route 43
    Departure Procedures (DPs)...................................43
    Standard Terminal Arrival Procedures (STARs) 47
    Anticipating the GPS Approach 47
    Weather Factors..51
    Choosing an Alternate 51
    Personal Minimums..52
    Filing the GPS Flight Plan 53
    Planning GPS Flights without DUATS....................53
        Black Box Practice—3 ....*55*

6 Departure, En Route, and Arrival 57
    GPS Cockpit Checks 58
    Copying Clearances 62
        Shortcuts That Don't Shortchange—5...............*63*
    Cockpit Confusion 63
    Before-Takeoff and Takeoff Checks 64
    Moving Map 67
    Distractions 67
    En Route Operations...68
    Messages 68
    VOR and GPS Differences 69
    An Airline Technique You Can Use 69
    En Route Clearance Changes.................................70
    Diversions 70
    Nearest Function 71

En Route Holding 71
VNAV Can Be Distracting 72
    Black Box Practice–4 ....*73*

7   GPS Approaches 75
GPS Overlay Approaches 75
GPS Stand-Alone Approaches 77
GPS RNAV Approaches with WAAS 77
New Approach Chart Items 81
Waypoint Symbols 84
RNAV Approach Minima 85
GPS Approach Basics 86
Setting Up the Approach 87
The Autopilot Advantage 88
Keystroke Briefings 88
Scaling 89
In the Cockpit on a GPS Approach 91
Approach Vectors 92
On Final 93
Missed Approach 94
    Black Box Practice-5 ....*95*

8   Outages, Emergencies, and Other Surprises 97
RAIM Warnings 97
Coping with a GPS Outage 98
RAIM Predictions 100
Other Emergencies 100
Two-Way Radio Communications Failure 101
Importance of Logging Times 101
Emergency Altitudes 101
Complete Electrical Failure 102
    Black Box Practice-6 ....*103*

9   Instrument Check Rides with GPS 105
GPS Changes Things 106
You Have GPS, but Is It Legal? 107
Flight Planning Review 107
Safe Altitudes 108
Oral Examination 108
Typical GPS Oral Questions 108
    Shortcuts That Don't Shortchange-6 ....*110*
Now Let's Test Your Flying Ability! 111
Common Deficiencies 112
    Black Box Practice-7 ....*114*

10 Looking Ahead 115
Transitioning to WAAS 115

GPS Costs 116
Glass Cockpits 116
Display Distractions 117
The Power of the PFD 118
   High-Tech Abbreviations and Acronyms *119*
What an MFD Can Show You 120
MFD Options 120
Airborne Weather Radar 121
Chart Viewing..121
Datalink Weather..121
Lightning Detection..122
Terrain Awareness 122
Traffic Advisories 122
Work in Progress 123
The Big Picture 125

11 GPS Flight Lesson Syllabus 127
Flight Lesson 1: Basic GPS Functions 127
Background Briefing 1–2: Background, Data Entry, and
   Flight Procedures...129
Flight Lesson 2: Departures and En Route GPS—Part I 130
Flight Lesson 3: Departures and En Route GPS—
   Part II 131
Background Briefing 3–4: GPS Approaches 132
Flight Lesson 4: GPS Instrument Approaches—Part I:
   Overlay and Stand-Alone Approaches.............................133
Flight Lesson 5: GPS Approaches—Part II: Approach
   Amendments and Missed Approach Vectors 135
Flight Lesson 6: GPS Approaches—Part III: Precision
   Approaches with WAAS..136
Flight Lesson 7: Long IFR Cross-Country with GPS 137
Background Briefing 7–8: Flight Checks With GPS 138
Flight lesson 8: Preparation for the GPS-Based FAA
   Instrument Check Ride 139

12 Reference Section 141
I. Sources and Resources 141
II. Selective Availability 142
III. DUATS Details 143
IV. Abbreviations and Acronyms....................................144
V. Glossary 149
VI. AIM Excerpts Relating to GPS 159
   1-1-20. Global Positioning System (GPS) *159*
   1-1-21. Wide Areas Augmentation System (WAAS) *181*
   5-4-5. Instrument Approach Procedure Charts *186*
VII. FAR 91.185 IFR Operations: Two-Way Radio
   Communications Failure 209

Index 211

INTRODUCTION

The global positioning system (GPS) has come upon us in general aviation like a tidal wave. We heard reports of it long before it reached us. We knew it was going to be big, but as pilots and flight instructors, we weren't quite sure what to make of it. And now that it has arrived, we must master it. It can't be ignored!

While it will be years before the current ground-based navigation is fully trimmed down to a secondary role in the airway system, GPS technology is now storming into the navigational forefront. For flight instructors, students, and experienced general aviation pilots, the arrival of GPS means that we must learn and effectively teach the new, space-based technology of GPS while staying sharp with the older ground-based technology centered around ILS, NDB, and VOR.

Added to this is the complexity of the new equipment coming on the market all the time. American avionics companies lead the world in the development of GPS receivers and displays. And, as is usually the case with new aviation technology, GPS equipment is becoming less expensive and more standardized all the time. But we're not there yet.

So, as flight instructors, we have had to ask ourselves, what is the most effective way to teach the new technology and make it a welcome partner in today's general aviation cockpits?

What techniques can we apply to make IFR pilots comfortable with the dual nature of their space-based–ground-based world? What can we do to clearly show how to use the basic functions of GPS when there is still much variation from one manufacturer to another in the ways these functions are configured?

We are deeply indebted to Henry Sollman for his pioneering work in the teaching of instrument flying. His innovative methods, which he shares so generously in his book *Mastering Instrument Flying,* have stood the tests of time and evolving technologies. He has graciously consented to the liberal use of material from his book in ours, and he is a general consultant on *Mastering GPS Flying.*

We have relied on a wealth of information from the FAA and other government agencies. The avionics industry has been encouraging from the beginning and has responded wonderfully to our requests for help. In Chap. 12, page 141, you will find a list of our main sources along with information on how to contact them.

This book was conceived over lunch in Oxford, Mississippi, one fine spring day when we were both living there, and we thank our families and friends for bearing with us over the long haul. Our thanks go especially to Joey Brent for his help in solving many computer problems and to Anna McGahey Sayre for her fine work in preparing all the illustrations for publication.

A book like this is never finished. We welcome your thoughts and comments.

2

MASTERING GPS FLYING
IN EIGHT LESSONS

So, are you new to flying on instruments? Or are you a veteran on the gauges? Either way, this book and its eight-lesson course will show you how GPS can take you wherever you want to go. And if you are VFR only, the flight lessons in this book can be adapted to make GPS a welcome partner in your cockpit. See Chap. 11: GPS Flight Lesson Syllabus, page 127, for details on the flight lessons and briefings you will work with in this course.

And remember, if you bring a GPS-equipped aircraft to an Instrument Rating Practical Test check ride, the examiner will probably ask you to demonstrate GPS approaches.

Flight Lessons

Flight Lessons 1, 2, and 3 will get you acquainted with the basics of GPS preflight planning, departures, en route procedures, and arrivals. These three flight lessons will provide the time you need to become familiar with the GPS keystrokes you will use to get to your destination ready to commence an instrument approach.

Together you and your instructor will review the necessary GPS cross-country navigation logs and flight plans. Then you will file IFR flight plans using real-time weather. Choosing an airport 40 to 50 miles away will keep the whole lesson under two hours in most general-aviation aircraft.

Instructors Note: Have your instrument students plan, file, and depart on an IFR flight plan on *all* training flights. And don't hesitate to carry out these flight lessons under actual IFR! This is the best possible practice—and you won't have to work with a hood in the clouds, nor will you be distracted by other traffic when you are under the control of ATC.

On the one hand, it is tempting to show off the wonders of GPS to a new student by going up without an IFR flight plan and dialing up a couple of "nearest" airports or going "direct-to" some nearby VORs, then demonstrating a couple of VFR approaches before heading back to the barn. But this doesn't really help your students all that much. On the other hand, if they file and fly IFR flight plans on *every* GPS training flight from the beginning, they will become competent very quickly in using the full range of GPS features in the ATC environment.

You can always cancel IFR and proceed to VFR to practice GPS routines such as diverting to avoid weather. If you do cancel IFR, contact ATC and request "flight following." ATC can be the safety net of another pair of eyes following you as you concentrate on GPS practice. But make it a practice *always* to depart and conduct your training flights on an IFR flight plan. Experience has shown that this technique produces more confident and proficient instrument pilots. And it helps eliminate the problem of poor planning—a big cause of failure on instrument rating check rides.

If your student is VFR only, have him or her plan a "GPS direct" VFR flight backed up by VORs. Watch out for restricted and special use airspace! Again, contact ATC for flight following as you concentrate on your VFR GPS training.

If the student has not yet obtained an instrument rating, all IFR flight plans must be filed in the instructor's name according to FAR 61.3 (e)(1). The nonrated student cannot act as pilot in command while flying on an IFR flight plan.

Prior to the IFR departure, you'll copy a clearance from ATC. Again, don't short-change yourself from the full learning experience. Both copying clearances and flying while coordinating with ATC give you real-world practice. If you just finished your private pilot training, you may still be shy about talking to a controller. Practicing by doing the real thing in a not-so-critical circumstance is a good way to learn. Copying clearances, coping with changes and amendments, and working with ATC on departure, en route, and approach procedures will become routine very quickly. This will enable you to spend more quality time honing your GPS proficiency.

When you're finally airborne, put on your hood and start building a proficient instrument scan, including the GPS display. During the en route phase, remember the golden rule of aviation: "aviate, navigate, and communicate." You must remain within 10° of your desired course, 10 knots of desired airspeed, and 100 feet of assigned altitude during all maneuvers. These are the tolerances specified in *Instrument Rating Practical Test Standards*. In later flights, you and your instructor will work on tightening up these parameters to a more professional "2, 2, and 20."

Don't worry about making instrument approaches on these first three flight lessons. Your instructor will set them up and demonstrate them at your destinations. When the instructor commences the approach, he or she should let you remove your hood and enjoy it. Follow along on the approach plate. Jot down your questions and save them for the postflight briefing.

Many instructors find that the learning atmosphere is more relaxed on Flight Lesson 1 if the first leg ends with a landing and a brief break. This gives students a chance to ask questions about GPS procedures they will practice on the way home.

Shortcuts That Don't Shortchange-1

Save yourself time—and money—on the day of the flight lesson by preparing your assigned flight plan at home a day or two ahead of time. This will allow you time to fully research all the departure procedures, waypoints, backup VORs, nearest and alternate airports, standard arrivals, approaches, and missed approaches you need to know for the assigned flight. Advance planning of this kind will give you a much better chance of avoiding surprises during the flight.

Also check the weather from home as you work on the flight plan or navigation log. Does it look like an alternate route will be required? If so, plan for it at home. Go ahead and check all NOTAMs while you have plenty of time. Be especially alert for Temporary Flight Restrictions (TFRs). TFRs have proliferated since 9/11. On the day of the flight call weather for an abbreviated briefing to update the weather before leaving for the airport. None of this type of advance planning will cost you a cent!

Flight Lessons 4, 5, and 6 concentrate on GPS approaches and their variations, including missed approaches. These lessons will prepare you for GPS overlay and stand-alone approaches and also introduce you to new RNAV approaches made possible by WAAS (wide area augmentation system).

Instructors Note: If the aircraft you will be using is not WAAS-certified, your student will not be able to practice LPV and LNAV/VNAV approaches because there will be no electronic glideslope capability. However, you can practice LNAV approaches.

From now on, you will make the approaches while the instructor observes and assists. Prepare, file, and fly IFR flight plans for all approach practice flights so that the transitions between the en route and approach phases of a GPS-based instrument flight are fully mastered along with the approach procedures themselves.

The instructor should have you fly down to minimums and execute a missed approach whenever circumstances permit. Missed approach procedures (MAPs) have proven to be a source of confusion on GPS-based approaches because they frequently require shifting out of GPS into the OBS mode for some portion of the MAP. You will find that the more you practice GPS missed approaches, the sooner they will become routine.

Flight Lessons 7 and 8 will give you additional opportunities to hone your skills with GPS in the real world of IFR. Your instructor may also use these flights to concentrate on fine points—GPS and otherwise—that could use some additional work. If needed, Flight Lesson 8 can also serve as the sign-off flight before certifying a student for an instrument rating flight test. Flight Lesson 8 can also serve as an Instrument Proficiency Check (IPC).

Tasks required for a proficiency check are listed in the *Instrument Rating Practical Test Standards* booklet with the letters PC next to each required task. All items labeled in this fashion should be completed for a pilot to be legally signed off for a six-month proficiency check.

Preflight and Postflight Briefings

You will see that the syllabus in Chap. 11, page 127, calls for preflight and postflight briefings on every flight lesson.

The preflight briefing will cover the objectives of the lesson, the procedures, maneuvers to be practiced, and the standards the student is expected to achieve.

The postflight briefing will clearly establish the elements of the lesson that the student has mastered, as well as those that need further attention. The postflight briefing will also include a preview of the next flight lesson and the material to be read or reviewed. And, because every lesson will start by filing an IFR flight plan, the instructor will suggest the destination(s) for planning and filing for the next lesson.

Students will progress much more rapidly (and at considerably less cost!) if they look up the assigned reading for each flight lesson and become familiar with it prior to the preflight briefing. Assigned reading describes in detail what will take place during each flight lesson. The reading material also contains information and background that must be understood to complete the flight lessons successfully.

Background Briefings

Refer again to the syllabus and note that there are "Background Briefings" prior to three flight lessons that introduce important new material:

Background Briefing 1–2: GPS background, data entry, and normal flight procedures. This briefing will give you and your instructor an opportunity to discuss topics such as how the satellite system is set up, the importance of RAIM (receiver autonomous integrity monitoring), how your specific GPS unit works, planning the GPS flight, and en route navigation by GPS.

Background Briefing 3–4: GPS approaches, including overlay, stand-alone, and WARS-based precision approaches. And it leads to discussions of the differences between GPS and non-GPS approach plates, how to fly vectored GPS approaches, what to anticipate on a GPS missed approach, and how to handle amended clearances. The focus in this briefing is on what to do to make a normal approach go smoothly.

Background Briefing 7–8: Flight checks with GPS. The final briefing covers the fine points of not-so-common situations that frequently show up on flight checks, such as holding with GPS (which has a couple of pitfalls), equipment failures, and partial panel and unusual attitudes. Students and instructors should consider the possibilities of a vacuum pump failure, a GPS unit failure, and lost communications procedures.

Students should prepare for background briefings by studying the assigned reading on their own and writing out answers to all the questions. There are no trick questions in the background briefings; however, it might take some research to find satisfactory answers. That is exactly what the briefings are meant to do—make you dig for the information yourself. What you learn by researching the material yourself will stick with you much longer than material learned by rote.

That's also the reason why no answers are supplied in this book. If the answers were readily available, there would be no incentive to dig out the information on your own. Your instructor will have the answers and will provide them when you discuss the background briefings.

As soon as you answer all the questions in one background briefing, start working immediately on the questions in the next briefing. You want to be prepared to go over the briefings with your instructor at the appropriate point in the syllabus, and it does take time to prepare all the answers.

Instructors Note: Be prepared! Be sure you have all the answers under control before you go over these briefings with your students. Or you may not be able to discuss all the questions and answers intelligently. You will have to locate some of the answers in other sources, such as GPS manuals and the *Aeronautical Information Manual* (AIM).

Do not confuse background briefings with preflight and postflight briefings. The latter are conducted at the beginning and end of every flight lesson. Background briefings are separate sessions between flight lessons. They contain too much material to be included at the beginning of a flight lesson.

Background briefings are numbered to show when they should be conducted. Thus Background Briefing 1–2 should be completed right after Flight Lesson 1 but before Flight Lesson 2.

The student and instructor should schedule background briefings as ground instruction time separately from flight lessons. The ground instruction sessions should last one to two hours and can be scheduled over two or three days. Students and instructors should go through the briefings together. Instructors should make certain that students clearly understand the material covered by the assigned reading and questions before proceeding with the next flight lesson.

Learning the Keystrokes

Newcomers to GPS are unanimous: The most difficult, frustrating challenge is learning how to enter and retrieve data without making mistakes or losing situational awareness.

New users also agree that absolutely the worst possible place to learn GPS keystrokes is in the cockpit of a single-engine plane in flight, which, of course, is where most of us do our instrument training and practice. It takes much longer to master the keystrokes while in flight, and it is almost impossible to get the keystrokes right without disrupting the instrument scan or making frequent, sometimes serious, errors.

Docking Stations

The best way to master the GPS keystrokes, we believe, is to practice them on a docking station. This device takes advantage of the easy removal of GPS units from the aircraft, a feature common to most newer general-aviation GPS receivers. With very little effort, the GPS unit can be unplugged from the aircraft, carried inside to a classroom, an avionics shop, or your home or office, then plugged back into the docking station for keystroke practice.

In this take-home mode, the docking station will allow you to practice data entry for every situation—including flight plan entries, departures, en route segments, arrivals, and approaches—as well as all the other functions of your GPS unit, such as nearest, direct-to, holding, missed approaches, and so forth. You can even "fly" your flight plan in real time. Or you can increase your TAS to 300 or 400 knots on long stretches and complete the practice flight more quickly.

For details and prices on docking stations, contact

Lone Star Aviation Corp.
604 South Wisteria Street
Mansfield, TX 76063-2423
Tel: 1-682-518-8882 or 1-817-633-6103
www.lonestaraviation.com

Before you sign up for GPS flight instruction, find out what system will be used to teach it (e.g., this system available to you when you need it. In other words, research and shop around for docking station availability as you would for a course, instructor, and rental airplane. There are other ways of practicing GPS keystrokes, and we will discuss some of these, but the docking station is by far the best.

And while we're talking about shopping around, be sure to look for an autopilot in the airplanes you might want to work with in your GPS training. Autopilots have moved from the "nice to have aboard" to the "absolutely essential" category with the arrival of GPS.

It is almost impossible to handle radar vectors, clearance changes, and all the many navigation chores that come up on an instrument flight without flying the airplane on autopilot while you are punching the GPS buttons—and repunching them if you make a mistake!

Especially in the beginning of your GPS training, becoming familiar with keystrokes takes repetition. This requires patience on your part as well as your instructor's. And you will, at times, hit a wrong key or get disoriented. Your instructor will teach you how to find a way to return to a familiar point on the GPS when you make a mistake. Using the autopilot in these situations will be an important factor in the successful management of your GPS cockpit resources.

If you are the instructor, you obviously need to practice the keystrokes yourself to a level of competency before you teach them, or you might teach something wrong. Remember the law of primacy: whatever is learned first is remembered best. If something is learned wrong the first time, you will then be testing this law to banish something learned wrong out of the student's mind!

Another good way to learn GPS keystrokes is to practice them in the cockpit of a GPS-equipped aircraft parked on the ground. But first, plug into external power before turning the electronics equipment on. Otherwise, you may run the battery down, especially in cold weather. If you don't have a plane of your own available, see if you can schedule the practice you need in one of your FBO's GPS-equipped planes connected to external power.

The only thing you won't be able to do in a parked airplane is simulate flying a flight plan. This is possible on the ground only when the GPS unit is disconnected from the aircraft's satellite antenna and operating in the take-home mode. If the unit is connected to the antenna, it will use satellite signals to determine position and movement. Since there is no movement when parked, there will be no movement indicated on the display.

Shortcuts That Don't Shortchange-2

Military pilots will tell you most emphatically that you don't ever want to be groping around the cockpit for the right knobs and switches in an operational or high-performance situation. To solve this problem, many years ago the military introduced a cockpit-familiarization technique called the "blindfold cockpit check." This taught pilots how to locate everything in the cockpit by touch rather than sight.

You really don't have to blindfold yourself to make the technique work well with any GPS setup or any other item in your cockpit. Just open and close your eyes as indicated below while running through the routines. But no cheating!

Step 1. With eyes open, touch each knob and button on your GPS, say its name, and learn the way it feels. When you have mastered this step, move on to step 2.

Step 2. Repeat step 1 with your eyes closed until you get all the items correct by touch alone. Then go to step 3.

Step 3. Close your eyes, and have someone else call out the various knobs and buttons at random while you locate them by touch.

When you can do step 3 with 100% accuracy, congratulations! You are now ready to twist the knobs and punch the buttons in flight!

Finally, check out the various videos and computer GPS simulators now available and see which ones can help familiarize you with the keystrokes for the equipment you'll be using in your training.

For example, ASA offers *GPS Trainer 2.0,* which covers six of the most popular GPS units on one CD-ROM. Look up ASA up on its Web site www.asa2fly.com or call 1-800-ASA2FLY.

Another example: Garmin offers an instructional video for the GNS 430 and GNS 530. Garmin also has a downloadable simulation program for the GNS 430 and GNS 530. Contact Garmin at www.garmin.com.

Black Box Practice

At the end of several chapters in this book you will find a special feature entitled Black Box Practice. These practice routines will help you master GPS keystrokes like a pro. Just look for this symbol:

To better simulate the real world of IFR as you practice, keep your eyes off the unit as much as possible while working on the practice routines. Locate the button or knob you want by feel, and set what you want by counting the clicks. Now look and verify or correct the setting and enter it.

Touch, set, verify, enter—with practice you will find that you can work the GPS display with a minimum of fixation on it, and a maximum of time for your normal instrument scan.

Don't Overlook This Fundamental

Regulations—and common sense—require a safety pilot in visual meteorological conditions (VMC) to watch out for other traffic, clouds, and obstacles such as radio towers and hills. According to FAR 91.109 (b), "no person may operate a civil aircraft in simulated instrument flight unless the other control seat is occupied by a safety pilot who possesses a private pilot certificate with category and class ratings appropriate to the aircraft being flown."

If you are acting as safety pilot in a Cessna 172, for example, your pilot certificate must have an airplane single-engine land rating; if you are in an Aztec, your pilot certificate must have an airplane multiengine land rating. A safety pilot must also have a current medical certificate, because the safety pilot is a "required pilot flight-crewmember" in terms of FAR 61.3 (c)(1).

3

ALL YOU NEED TO KNOW ABOUT GPS TECHNOLOGY

Let's say you are really serious about understanding GPS, you are studying hard, and you read this sentence from a basic GPS *Advisory Circular* (AC):

Equipment error assumes an average HDOP of 1.5, GPS equipment waypoint input resolution of 0.01 minute, and coordinate output resolution of 0.01 minute for approach and 0.1 minute otherwise.

How do you handle technological overkill like this? Ignore it? No—what if it's something you need to know? Keep on reading? Ask your flight instructor?

It's a dilemma. We pilots basically like high-tech. We enjoy getting immersed in it, and we want the latest and best for our cockpits. But in the case of GPS, "total immersion" in the technology itself doesn't necessarily translate into a confident, comfortable skill in using GPS. Total immersion can even be counterproductive when it comes to learning the nuts and bolts of how your equipment works.

The dilemma is even more acute for renter pilots because of the lack of standardization among competing brands of GPS cockpit systems. As you encounter the different systems while hopping from one plane to another, the temptation is to learn the direct-to function and forget about the many other important features of this revolutionary technology. This could become a serious handicap if you hope to move on up the ladder into a corporate or airline job.

In this chapter we present all the technology you need to understand how GPS works and how to make it work for you.

Don't "Read" the Manuals!

With GPS manuals running to more than 300 pages, it is extremely frustrating to learn how to use a specific GPS system by reading the manual from cover to cover.

Instead, use the manual as a reference—a miniencyclopedia for looking up the details of something specific, such as VNAV (vertical navigation). Current manuals are excellent when used this way.

As you approach the challenge of learning how to use GPS comfortably and effectively, think in terms of cutting to the bare technology essentials that you *must* know to work with GPS on an everyday basis. Learn to use the *must* features first and put the background details of technology aside for a rainy day when no one is going anywhere.

If you wish to proceed deeper into GPS technology after you finish this chapter, you will find good, thorough wrap-ups in the *Aeronautical Information Manual* (AIM). For your convenience, the relevant GPS sections of AIM—including the current information on WAAS, the Wide Area Augmentation System—are produced in full in the Reference Section (Chap. 12) of this book starting on page 141.

How GPS Works

With the arrival of GPS, we are poised to phase out ground-based systems, such as VOR, NDB, ILS, LORAN, and the like, and replace them with a space-based satellite system. This transition ultimately will have major implications for all of aviation. But for now, what do you really need to know about GPS technology to begin using it?

The basic concept is not all that complicated. GPS has three major elements: (1) a constellation of 24 orbiting navigation satellites, (2) a ground-based network consisting of one master control station and five monitoring stations, and (3) a wide variety of receivers for air, land, and sea applications.

The 24 satellites are in orbit 10,898 nautical miles above the Earth. (See Fig. 3.1.) This configuration provides complete global coverage 24 hours a day for an unlimited number of aircraft. As the satellites move along their orbits, each one continuously transmits a stream of precise data on its changing position. Each satellite also transmits a stream of time signals which are, in effect, continuously "time stamped" to indicate the exact moment the signals leave the satellite.

The satellites are monitored by the second element of GPS: a ground-based network of five monitoring stations and one control station. The five monitoring stations are located in Hawaii; on Kwajalein, Diego Garcia, and Ascension islands; and at Falcon AFB, Colorado Springs, Colorado. The one control station is also located at Falcon AFB.

The monitoring stations continuously collect electronic data from the 24 satellites and forward the data to the control station. Anomalies or errors in the satellite signals are analyzed at the control station. Corrections generated by this analysis are transmitted back from the control station to the satellites so they can adjust their outgoing signals.

The third and final elements of the GPS system are the units that receive the satellite transmissions and use them to compute and display position and navigation data.

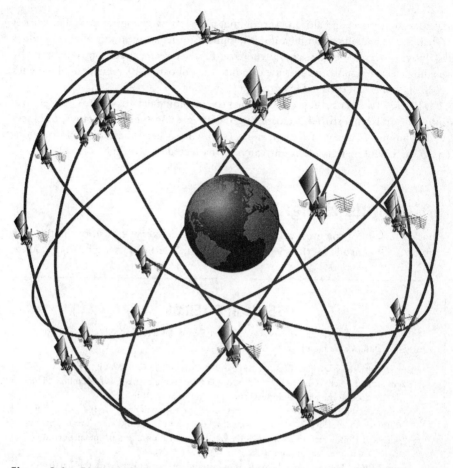

Figure 3.1 FAA artist's conception of GPS satellite constellation. Each satellite completes one orbit every 12 hours.

For aviation use, they come in many shapes and capabilities these days, from small handheld units the size of a cell phone to big, multifunction displays (MFDs) found in the glass cockpits of our most modern jets.

When streams of GPS time and position data arrive at an aviation GPS receiver, they are processed by a computer in the GPS receiver. The arrival times of incoming time signals are compared to their "time-stamped" times of departure from the satellite. Time differences are then converted into distances. These distances are correlated with incoming position signals to pinpoint the aircraft's position on a line between the aircraft and the satellite. Four such lines between an aircraft and a minimum of four different satellites provide a three-dimensional position consisting of latitude, longitude, and altitude throughout a constantly changing time. For greater reliability, the satellite constellation is designed so that a minimum of five satellites is always observable by a user anywhere on Earth. Greater precision is obtained by feeding electronic pressure altitude data into the onboard GPS computer, along with correct altimeter settings. This is called *baro-aiding*.

The GPS receiver's computer combines the incoming data from the satellites with information from a stored database to compute ground speeds, bearings, distances to waypoints, estimated times en route, and many other functions. Transmissions from the satellites are made at a frequency unaffected by weather.

Great Circle Navigation

Another reason for the great accuracy of GPS is its delivery of positions in terms of latitude and longitude on the surface of the Earth. The Earth is a sphere, of course,

---

USEFUL GPS TERMS

Some new terms have come into growing use of with the arrival of GPS. Here are some you will encounter frequently:

Baro. Altimeter setting. You "set the baro" as part of the GPS preflight checks.

Baro-aiding Supplying pressure altitude data to the onboard GPS computer. This is like adding the altitude inputs from an additional satellite.

K First letter of the international identifier for most airports in the contiguous United States; KLAX for Los Angeles, for example. C indicates an airport in Canada; P indicates an airport in Alaska. If there are numbers in the airport identifier, these prefixes are not used.

Track (TK) The actual flight path of an aircraft over the surface of the Earth, not the same as course, which is the intended direction of flight measured in degrees from north. Track is a great circle term; course is a flat map term.

Desired track (DTK) The planned or intended track between two waypoints.

RNAV Umbrella term for enhanced navigation systems that can compute position, track, and ground speed, then provide distance, time, and cross-track error to the pilot. Present day RNAV includes INS, LORAN, VOR/DME, and GPS systems.

Waypoint A predetermined geographical position used for route and instrument approach definition, progress reports, visual reporting points or points for transitioning or circumnavigating controlled or special use airspace. Waypoints are defined relative to a VORTAC station or in terms of latitude and longitude coordinates.

User-defined waypoints These are handy customized waypoints that you can set up with GPS to locate private airports not included in your database, go back to agricultural fields sprayed regularly, return to good locations for practicing ground reference maneuvers, and so forth.

so GPS positions are real-world positions on a curved, global surface. GPS courses and distances are not derived from something that has been projected and distorted to produce the convenience of the flat maps we normally work with. Instead, GPS courses and distances are based on "great circle" courses and distances, which are the shortest distances between two points on the curving surface of the Earth (or any other globe).

For short flights the differences between great circle and flat map headings and distances are small—often negligible. But on long hauls, such as transoceanic and transcontinental flights, the differences become noticeable, and they must be reckoned with for efficient flight operations. (See Fig. 3.2.) Fortunately, GPS automatically gives us great circle solutions to all air navigation problems. In Chap. 5, How to Plan a GPS Flight, we will show you how this works in the real world of IFR.

VFR Use of GPS

For around $1,500 you can now buy handheld, VFR-only GPS receivers (as advertised in a recent *Sporty's Pilot Shop* catalog) that include a Jeppesen database "with airports, VORs, NDBs, intersections, special use and controlled airspace, runway

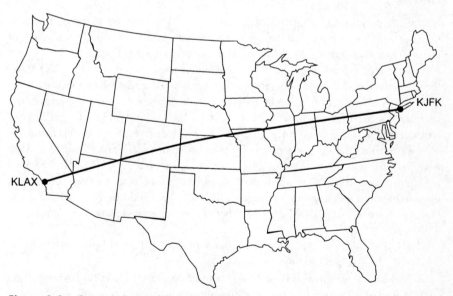

Figure 3.2   Great circle route from John F. Kennedy International to Los Angeles International. No-wind headings change from 288° on departure from KJFK to 232° approaching KLAX. The great circle route is 19.5 nm shorter, as computed by DUATS.

data, communications frequencies and final approach sequence waypoints." The database also depicts "lakes, rivers, railroads, state/national boundaries, and coastlines." And all this is in color!

With all these features, why are these receivers VFR-only? There are five principal reasons: (1) certification, (2) power source, (3) database currency, (4) antenna installation, and (5) integrity monitoring.

There are two necessary certifications for IFR use. First, the manufacturer of the unit must meet the exacting standards of FAA's Technical Standard Order TSO C 129, TSO C 145a, or TSO C 146a for the unit to be legal for IFR use. TSOs C145a and C146a involve the augmentation of GPS system with WAAS capability.

All panel-mounted GPS units now comply with TSO C-129. No handheld or yoke-mounted units meet the requirements for IFR certification under TSO C-129.

Second, GPS units must be *installed* in accordance with the FAA's Advisory Circular AC 20-138. To qualify for IFR under AC 20-138, a GPS unit basically must have (1) a permanent connection with the aircraft's main power source, (2) provision for an updatable database, (3) a permanently installed antenna that provides a clear, uninterrupted view of all satellites above the horizon, and, last but not least, (4) integrity monitoring, or RAIM.

RAIM

Receiver autonomous integrity monitoring, or RAIM, is an independent function within the GPS receiver that continuously monitors and verifies the integrity of incoming satellite signals. RAIM warns if data from any satellites are missing or corrupted too much to be reliable. Unlike the warning flags that pop up when VOR and ILS are unusable, the RAIM operation is independent of ground-based facilities.

The RAIM concept is a simple one. RAIM monitors one satellite more than the number required for accurate GPS navigation For example, if four satellites are required for an accurate GPS fix, RAIM will monitor five. If the signals from one out of the five satellites are substantially inconsistent with signals from the other four—or the required number of satellites cannot be found—RAIM will indicate a problem. The message annunciator (M or MSG) will light up to alert you that an important message has come through. This message should be read immediately. Here are some typical RAIM messages that run on three widely used systems:

RAIM *Position Error.* There is a problem with one of the satellites
   (Bendix/King KLN 94).
RAIM *not available from FAF to MAP waypoints.* Sufficient satellite coverage does
   not exist for protection limits of this approach segment (Garmin GNS 530).
En Route GPS RAIM Not Available. RAIM is not available for en route phase
   (Apollo GX60).

The limitations on the IFR use of handheld and yoke-mounted GPS units all make good sense.

You must have a current database for all IFR work. This requires a database that can be updated frequently as changes are published. Handheld systems do not usually have this capability.

As for the permanent power source and the permanently installed antenna, you don't want to lose contact with your satellites in the soup when your batteries die, or when you make a turn that blanks out your handheld or yoke-mounted GPS antenna.

Above all, you need RAIM to guarantee a sufficient number of healthy satellites for a safe, efficient IFR flight. Handheld and yoke-mounted GPS units do not provide RAIM.

How Accurate Is GPS?

The signals stream in from the satellites at the speed of light, 186,000 miles per second. To be accurate at this speed, time must be measured in nanoseconds, or *billionths* of a second. The technology needed to process a stream of data moving at nanosecond speed is perhaps the most impressive achievement of GPS and the main reason for its accuracy.

How accurate is GPS exactly?

For years, GPS's military managers operated GPS under a policy of *selective availability*. GPS signals were deliberately distorted to prevent civilian equipment from being used for pinpoint accuracy against American targets. The civilian level of service was known as standard positioning service (SPS), and it was accurate to about 100 meters (328 feet) or better.

Shortcuts That Don't Shortchange-3

In spite of its limitations, VFR-only GPS can be a very helpful cross-check on a VFR flight. The FAA encourages this use. "GPS navigation has become a great asset to VFR pilots, providing increased navigation capability and enhanced situational awareness, while reducing operating costs due to greater ease in flying direct routes," according to AIM (see page 161). "While GPS has many benefits to the VFR pilot, care must be exercised to ensure that system capabilities are not exceeded. . . . VFR pilots should never rely solely on one system of navigation. GPS navigation must be integrated with other forms of electronic navigation (when possible) as well as pilotage and dead reckoning. Only through the integration of these techniques can the VFR pilot ensure accuracy in navigation."

The military level of service is known as precise positioning service (PPS), and it is accurate to about 40 feet or better. At least one FAA source has described the military signals as having "a margin of error of only a few centimeters."

On May 1, 2000, President Clinton signed an order ending selective availability as part of an ongoing effort to make GPS more attractive to civil and commercial users worldwide. Now, the GPS that civilians use is accurate to the military standard of 40 feet or better. The Interagency GPS Executive Board (IGEB), which governs the GPS system, announced that the United States has no intent to ever use selective availability again. See *"Selective Availability"* in Chap. 12, Reference Section, pages 142–143, for the full FAA statement on GPS accuracy.

This accuracy is sufficient for the use of GPS for IFR operations during departure, en route, arrival, and published GPS nonprecision approaches. And GPS has been approved for these uses. But this is not good enough to replace precision approaches such as ILS.

To provide the improved degree of accuracy for ILS, the FAA is committed to two major improvements to GPS: the Wide Area Augmentation System (WAAS) and, further down the line, the Local Area Augmentation System (LAAS).

How WAAS Works

Like GPS itself, WAAS has three basic elements: satellites, ground stations, and receivers.

The transmitting WAAS satellite constellation consists of two GEO (geostationary Earth orbiting) satellites in precise, fixed positions high above the Atlantic and Pacific oceans. A third GEO is planned for stationing over the United States. It will provide redundancy for the other two and fill some holes out toward the edges of the current coverage.

On the ground there is a network of 25 precisely surveyed WAAS ground reference stations (WRSs), two wide area master stations (WMSs), and four ground uplink stations (GUSs).

Signals from the GPS satellites passing overhead are continuously monitored by each WRS to determine the differences between the location of the ground station according to the passing GPS satellites and where the ground station really is according to its precise, surveyed benchmark location.

Each WRS relays its satellite data to one of the two WMSs where correction information is computed for specific geographical areas. Correction messages are prepared and uplinked via GUSs to a GEO. Corrected messages are broadcast by the GEO on GPS frequencies to WAAS-equipped GPS receivers across the United States and portions of Alaska.

Because WAAS measures the *difference* between where the GPS satellites say they are and where they really are, WAAS is sometimes referred to as *differential* GPS or DGPS. (LAAS is based on a similar concept and is also a type of DGPS.)

In addition to providing correction signals, WAAS GEO satellites also originate streams of GPS time and position signals of their own, providing, in effect, an additional GPS satellite that is always in view throughout the coverage area. This adds to the accuracy and reliability of the overall system.

If a WAAS receiver is "WAAS certified," it receives corrected GPS signals automatically and continuously. The system is seamless and invisible; there is nothing you have to do to get the corrected signals. If you are operating with a WAAS-certified receiver, you are working with corrected position information as soon as the GPS master switch is turned on.

WAAS became operational for WAAS-certified GPS receivers at 12:01 A.M. on July 10, 2003. The WAAS network of satellites and ground stations now provide coverage for the eastern and western United States plus lower Alaska. "The FAA is also working with Canada and Mexico to expand the WAAS coverage area to support North American implementation of WAAS," according to a recent FAA fact sheet.

WAAS is now being used by commercial and general aviation to provide precise guidance to runways that don't merit the cost of an ILS. WAAS will tweak the accuracy of GPS enough to allow approaches down to 200 feet AGL and ½ mile visibility at even the most remote airfields.

However, LAAS is primarily being developed with airliner accuracy in mind. These local systems have the potential for "zero/zero" landings. They might even be able to provide precise positioning for taxiing aircraft on socked-in, foggy taxiways. Can you believe taxiing on instruments?

LAAS Progress

Local Area Augmentation Systems will be based on ground stations at the airports they serve. LAAS, like WAAS, "augments" GPS by making continuous corrections to the basic streams of GPS data flowing from the GPS satellites.

LAAS will provide local corrections to GPS position information with accuracy comparable to ILS Category II and III approaches, or better. One LAAS station can provide runway guidance for several runway ends. Of great interest to the general aviation community is the possibility that LAAS can serve some airports where ILS can't be located because of space or radio frequency spectrum constraints.

The first LAAS systems are scheduled to be installed at Juneau, Alaska; Phoenix, Arizona; Chicago, Illinois; Memphis, Tennessee; Houston, Texas; and Seattle, Washington. The first LAAS system is scheduled to be operational by late 2006.

Approaches

In the past few years, hundreds of *nonprecision* GPS approaches have been approved for airports large and small all over the country and throughout the world. There are two types of nonprecision GPS approaches: (1) the GPS overlay procedure, in which the GPS procedure is laid down over an existing VOR or NDB procedure, and (2) the GPS stand-alone procedure, which does not require inputs from ground facilities such as VOR, NDB, or DME.

The big advance of WAAS over unaugmented GPS is that WARS is capable of delivering the vertical accuracy needed to generate electronic glide paths similar to ILS approaches. The first seven WAAS approach charts were published in NOS booklets on December 25, 2003.

Approach charts for WAAS precision approaches are now being issued on a regular basis. As it says in AIM, "WAAS will allow GPS to be used as the aviation navigation system from takeoff through Category I precision approach when it is complete."

Lower Minima

How can you tell if an airport has WAAS approaches? Look for airports with RNAV GPS approaches. Charts with LPV approaches heading up the "minima" box have WAAS-approach capability. (See Fig. 3.3.) For more details, see also "Terms/Landing Minima Data," on page 1 of any NOS *U.S. Terminal Procedures* booklet. As always, DA (decision altitude) indicates *a precision* approach; MDA (minimum descent altitude) indicates *a nonprecision* approach, as shown below:

1. LPV DA (localizer precision with vertical guidance)
2. LNAV/VNAV DA (lateral navigation with vertical navigation guidance)
3. LNAV MDA (lateral navigation)

The Local Area Augmentation System, LAAS, promises GPS precision approaches with minimums comparable to ILS Category II and III. The FAA is officially committed to the development of LAAS, but full implementation is still a few years away.

Chapter 7 will describe how to fly GPS approaches in detail. Meanwhile, as you plan and prepare for approaches in the lessons ahead, you should consider the following situations or scenarios, all of which GPS is designed to guide you through:

1. Transition from a STAR (standard terminal arrival) to a normal GPS nonprecision approach
2. Entering and departing holding patterns

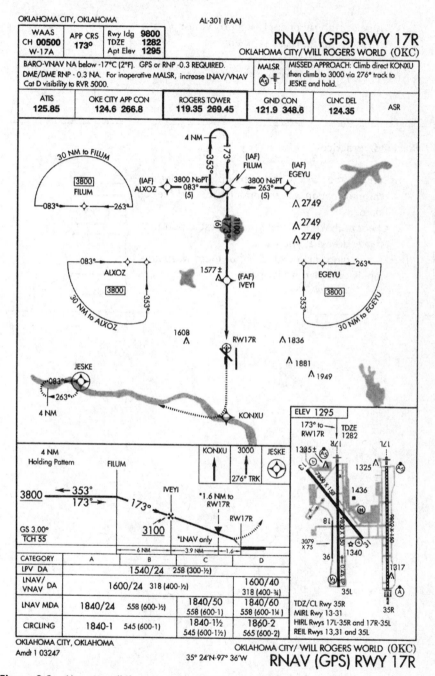

Figure 3.3 You can tell if an airport has WAAS approaches if it has RNAV GPS approach charts with LPV minima, as shown on this chart for Oklahoma City.

3. Change of runway or approach procedure
4. Vectors to final approach segment
5. Missed approach and proceed to alternate
6. Missed approach and return for another pass

Black Box Practice-1

1. Turn on your GPS. Check start-up screens for correct information. Enter correct baro and fuel on board.
2. Check to make sure that your nearest airport preferences are what you want.
3. Enter a flight plan with three or more waypoints, store it, and then retrieve it.

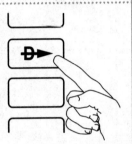

4

BUTTONS, KNOBS, AND SWITCHES

For most of us, our first contact with a GPS cockpit unit was probably confusing, frustrating, intimidating—or all of the above. There seem to be too many buttons, knobs, and switches—and some of them have more than one function! There are too many options and defaults to contend with. Really, why do we have to worry about the time zone in American Samoa? Punch the wrong button and all seems to be lost—not a good feeling on a missed approach!

Fortunately, you are not alone in this situation. It's something we've all experienced. And experienced GPS users have worked out many ways of simplifying GPS operation. By now techniques have evolved that make sense and allow for comfortable, convenient, and safe use of GPS under all flight situations. The techniques we are about to discuss are valid for the three most common lines of general aviation GPS units: Bendix/King, Garmin, and Apollo. See Figs.4.1 to 4.3.

As this was written, UPS Apollo was acquired by Garmin. Since there are many Apollo GX60 systems in use in training aircraft, we include it in this discussion. For further information on the GS60 and the rest of the Apollo line, contact Garmin, as listed the Chap. 12, Reference Section, pages 141–142.

Forget American Samoa!

Since GPS is designed for military use anywhere in the world under a variety of conditions, its programs and databases contain a wealth of worldwide geographical information. You may never need a lot of this information in routine IFR operations in the United States. Forget about American Samoa! Don't worry about whether or not WGS-84 is the correct Earth model for you (it is). If you should wake up one fine morning and find yourself in American Samoa, you can always look up the time zone in your hefty manual.

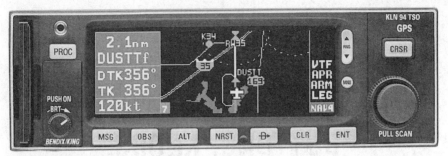

Figure 4.1   Bendix/King KLN 94. *(©2004 Honeywell International, Inc. All Rights Reserved.)*

Figure 4.2   Garmin GNS 530. (Courtesy of Garmin Ltd.)

It helps to approach the GPS cockpit unit as you would any desktop computer, because that's really what it is. Like a computer, the cockpit unit is a piece of hardware that provides the same basic functions regardless of who makes it. The flight software programs that it runs are pretty much the same from one unit to another, with variations in terminology from one manufacturer to another, just as Microsoft Word and Corel WordPerfect provide similar word processing functions, with different terminology and keystrokes.

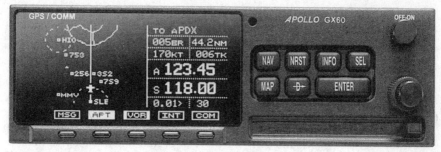

Figure 4.3 Apollo GS60. (*Courtesy of Garmin Ltd.*)

So what are the major GPS functions that we must master to use the equipment with ease and confidence? They are surprisingly few in number when you come right down to it. Let's take a look at these major functions in the order you are most likely to encounter them as you plan and fly an IFR flight.

Power Up

Before you turn on a new or unfamiliar GPS unit, you should check out some important basic interfaces. There is less standardization on these interface items than on any other features of GPS:

1. *NAV/COM.* Are your navigation and communications functions integrated with your GPS unit? If so, GPS will display frequencies for you to activate. If not, enter the frequencies manually, just as you always have. Be sure you know the procedures for setting up and switching NAV/COM frequencies for the GPS unit you are about to use.
2. *GPS/NAV.* How do you switch back and forth between GPS and your basic NAV functions, VOR and ILS? Is there an annunciator light that indicates when you are navigating in the GPS or the NAV mode (sometimes referred to as VOR/LOC)? You must be prepared at all times to switch out of GPS and into NAV if there is a problem with the GPS or if you must switch to ILS at your destination because of lowering minimums. The most common way of switching back and forth is by using the GPS/NAV button or toggle.
3. *Moving-map accuracy.* Is the moving map controlled separately from the GPS display? What ranges and distances can it show? What do the colors and symbols indicate? What options can you select, such as north up or track up orientation?

Now turn on the power to your GPS computer. The first minute or so will tell you some important things about the status of your equipment, regardless of make or model.

The first page will display a message from the manufacturer while the unit conducts a self-test. Subsequent pages show database expiration date and the results of the unit self-test. You will be asked to acknowledge this information. You will also be asked to enter the local altimeter setting ("baro") and the fuel on board.

Once you get beyond these opening pages, your GPS unit will begin to behave more and more like a computer. GPS has overall programs—or functions—to choose from. Within these functions there are options—or pages—you can choose as you need them. Like a computer, there is a cursor to get you where you want to go within a selected page. And there is an ENTER button to tell GPS what to do or to confirm a choice. These basic functions are the same for the Bendix/King KLN 94, Garmin GNS 530, and Apollo GX60 (shown in Figs.4.1 to and 4.3).

Unfortunately, standardization becomes scarcer and scarcer as you try to master the major functions of your GPS. See Table 4.1 to see how complicated this can get. Note that "direct-to" is the only major function accessed the same way by all three systems.

If at all possible, avoid switching from one manufacturer's system to another until you have mastered one system completely, to the point where you can use it confidently and with a minimum of distraction.

Do-It-Yourself Checklists

By now you may already be confused if you are a first-time user. It's so easy to get hung up as the unfamiliar choices pop up. Fortunately, there is a simple solution for

TABLE 4.1   Where to Access Major GPS Functions

	Bendix/King KLN 94	Garmin GNS 530	Apollo GA60
Direct-to	➡ button	➡ button	➡ button
Nearest	NRST button	NRST pages	NRST button
Enter flight plan	FPL pages	FPL button	Create a new flight plan page
Fly flight plan	FPL 0 page	FPL button	Active flight plan page
Enable turn anticipation	SET 10 page	(Automatic)	(Automatic)
Make an approach	PROC button	PROC button	FPL or NAV button
Vertical navigation (VNAV)	ALT 2 page	VNAV button	NAV button
Make a missed approach	➡ button	OBS button	OBS/HOLD annunciator

this problem, a solution developed by experienced GPS users who—like you—were also first-time users at some point. The solution is to write out your own do-it-yourself checklists to help you nail whatever is confusing or hard to remember. These do-it-yourself checklists do not have to be long or elaborate.

For example, a simple power-on checklist for a Garmin 530 might read something like this:

POWER ON

1. Power on: COM power knob clockwise
2. Database current? Press ok
3. Self-test indications:
   i. CDI: half left/no flag
   ii. TO/FROM flag: to
   iii. Bearing to destination: 135°
   iv. Distance to destination: 10.0 nautical miles
   v. All external annunciators (if installed): on
   vi. Glideslope: half up/no flag
   vii. Time to destination: 4 minutes
   viii. Desired track: 149.5°
   ix. Ground speed: 150 knots
4. Set full fuel? ok (or set actual FOB)
5. To set FOB:
   i. Rotate large right knob to select FOB.
   ii. Rotate small and large right knobs to enter desired figure.
   iii. Press ENT.

The beauty of do-it-yourself checklists is that you can add to them or subtract from them—or stop using them altogether—as your proficiency improves with experience. You don't even need to print these checklists out. Write them out on cards by hand—or shorthand—if you wish. If you need to learn a second or third GPS system, you can easily rewrite your existing checklists to work with the new systems since the basic functions are the same.

Direct-To

GPS carries out most of its functions in a direct-to mode. With GPS you do not track outbound from a VOR or an NDB. Instead of "to-from" when crossing a way-point, it's always "to-to." You fly to a waypoint, cross it, then proceed direct-to the next waypoint no matter how far away it might be. In theory, you can depart KJFK and fly GPS direct to KLAX, although, in the real world of IFR, ATC may feel more comfortable routing you via some en route waypoints.

You can enter a direct-to waypoint on the ground or in the air, VFR or IFR. Simply press 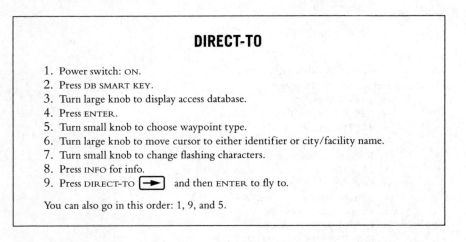, specify the identifier for the waypoint you want, and then enter or activate what you have specified.

The do-it-yourself checklist for direct-to can be very simple. Here is one for the Apollo GX60 suggested by Alexis Graves, a CFI from northern Mississippi, who flies with the CAP (Civil Air Patrol):

DIRECT-TO

1. Power switch: ON.
2. Press DB SMART KEY.
3. Turn large knob to display access database.
4. Press ENTER.
5. Turn small knob to choose waypoint type.
6. Turn large knob to move cursor to either identifier or city/facility name.
7. Turn small knob to change flashing characters.
8. Press INFO for info.
9. Press DIRECT-TO ⭢ and then ENTER to fly to.

You can also go in this order: 1, 9, and 5.

There is one very important exception when GPS does *not* operate in a direct-to mode. And that is when the OBS mode is in effect.

OBS Mode

Normally GPS operates in the "Leg" mode with automatic waypoint sequencing. But sometimes it is desirable to suspend automatic waypoint sequencing to comply with radar vectors, holding patterns, and procedure turns. To suspend automatic waypoint sequencing select the OBS function. The CDI/HSI will then emulate VOR course guidance. You'll still be hooked up to the GPS, but you will be able to set in your own courses and get VOR-like "to" and "from" indications. When your maneuvering is done, deselect the OBS function, and you will return to the Leg mode.

Entering Flight Plans

As we stress throughout this book, you should always plan and file an IFR flight plan for every Flight Lesson. Not only will you learn more quickly (and save money) if you do, but you will be better prepared for the real world of GPS flying. To use GPS comfortably and effectively you must carry a piece of paper into the cockpit that has all the numbers to punch in to get you where you want to go. You can't wing it and expect good results! And, as we shall see, there are special features available when you enter a flight plan that will make your work much easier.

Again, the basic flight plan function is the same across a wide variety of different manufacturers and models, although the terminology and method of accessing the function will vary from one system to another. With the Bendix/King KLN 94, Garmin GNS 530, and Apollo GX60, for example, you enter your flight plans on flight plan (FPL) pages. But entering the specific details is done differently on all three systems. Here is a sample do-it-yourself checklist for creating a flight plan with the Bendix/King KLN 94:

CREATING A FLIGHT PLAN WITH THE BENDIX/KING KLN 94

1. Select FPL 0 with the right outer knob.
2. Turn on the cursor with CRSR button.
3. Use the right inner knob to select the first character of the departure waypoint (K for most U.S. airports).
4. Repeat step 3 as needed to select the full identifier for the departure waypoint.
5. Press ENT to display waypoint page for identifier. If incorrect, press CLR and begin again.
6. Press ENT again to approve waypoint being displayed.
7. Repeat procedure to enter the rest of the waypoints.
8. Rotate right outer knob to move cursor manually to check that all the waypoints are entered correctly.
9. To activate the flight plan, rotate the outer knob counterclockwise to position cursor over USE? and press ENTER.

Waypoint Autosequencing

One of the great benefits of GPS becomes available to you when you create and fly a flight plan with two or more waypoints. This benefit is the automatic sequencing of waypoints. As you proceed along the string of waypoints in your flight plan, your GPS unit will automatically display the next waypoint ahead and count down the time and distance to it, as well as provide headings that will keep you on your desired great circle track. Autosequencing continues as you transition from the en route phase into an approach procedure.

Waypoint Alerting and Turn Anticipation

Shortly before you reach a waypoint in your flight plan, an annunciator light or message will alert you that the waypoint is coming up. If a turn of more than a few degrees

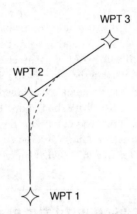

WPT 3

WPT 2

WPT 1

Figure 4.4 The turn anticipation feature of GPS eliminates bracketing during turns at a waypoint.

is required to put you on track for the next waypoint, a neat GPS feature called *turn anticipation* kicks in, and your CDI will lead you through the turn. (See Fig. 4.4.) In some systems turn anticipation is not automatically enabled. If this is the case, turn it on during your GPS preflight check. It's a big help in reducing your workload because it eliminates the need to bracket back and forth to steady up on your new track.

Frequency Autoselection

One helpful GPS option that has almost become a necessity is the combined NAV/COM and GPS cockpit unit. In an integrated system, each time you select a waypoint on your GPS, the database also selects a menu of NAV/COM frequencies you might need to work with at that waypoint. With a minimum of moves you can highlight the frequency you want to use, move it to STANDBY, and then toggle it into ACTIVE when ATC tells you to change frequencies.

In other words, the frequencies are all preset. They can be sequenced in the order you want to use them. With a minimum of fumbling around you can quickly and confidently move through an IFR sequence such as ATIS, clearance delivery, ground control, tower, and departure control. No need to write out long lists of frequencies on your flight plan! They're already in the database.

Nearest Function

Let's say you are assigned an altitude for a long IFR flight, and the headwinds are much stronger than you planned for. You run a fuel check, and see that you won't have enough to make a missed approach at your destination and proceed to your best alternate. Or you're in IMC at night, and you begin to pick up some serious ice. Or there is an impenetrable wall of thunderstorms up ahead. Or your engine is beginning to run rough.

You need to get down as soon as possible. Where is the nearest airport?

Fortunately, this is an easy one with GPS. Simply go to the NRST function, select the page for airports, scroll down the list, choose the one that best matches your circumstances, and then go there [→].

If that wall of thunderstorms is getting closer, pick an airport behind you that you can get into. If you are in IMC, you might want to select an airport expected to be VFR when you arrive. ATC can help with this. The NRST function can also supply you with nearest information on VORs, NDBs, ARTCCs, FSS locations, and other useful items.

Students and Instructors Note—Nearest Airport Defaults: As part of your GPS preflight check, be sure to verify that the defaults for nearest airport runway length and runway surface are what you want. As the Garmin 530 manual puts it, "You may wish . . . to exclude seaplane bases or runway lengths which would be difficult or impossible to land upon."

Vertical Navigation (VNAV)

Climbs and descents at 500 to 1,000 feet per minute typically work fine for light aircraft. But when do you start down to reach pattern altitude at the arrival airport?

The VNAV, or vertical navigation feature, allows the pilot to set up the GPS unit so that it will issue an alert when it is time to start a descent (or climb) to reach a desired altitude. The pilot enters parameters such as starting altitude, vertical speed in feet per minute, level-off altitude, and level-off distance in miles from the destination, if not the destination itself.

The box does the rest. An alarm signals or the message annunciator flashes when it is time to begin descent based on your settings and the aircraft's current ground speed.

The VNAV feature of GPS also automatically computes and displays rates of descent or climb when you set in the target altitude you're shooting for, your anticipated ground speed, and the distance to level-off. It's a big help to know what fpm to shoot for to avoid excessive air speed or overshooting your altitude when descending from, say, 18,000 feet as you approach your destination.

Up or down at 500 to 1,000 feet per minute usually works just fine for light aircraft. But as you move into heavier and faster planes, you will need greater precision than this to manage vertical speeds. Changing altitudes takes much more planning and attention in a King Air than in a Cessna172.

There are two things you should remember when using VNAV. One, set the unit up as early as you can when your workload is light. Also, a slower aircraft is fine descending between 500 and 1,000 fpm. But a faster plane may require between 1,500 and 2,500 fpm. When it's your turn in the left seat of something faster, practice first with different descent rates and VNAV inputs on a clear day; VNAV is a luxury that shouldn't detract from your aviating.

Oops! I Punched the Wrong Button!

Okay, okay—don't panic. Note that there are no panic buttons on the faceplate of your GPS unit. Instead, you have to return to a previous point where everything was set up the way you wanted it. Then you can resume what you were doing and try to reset the GPS correctly.

Hitting the wrong button on the ground is frustrating enough, but in the air, this mostly helpful little black box can consume enough of your attention that you forget about traffic, ATC, terrain, your other instruments.

A recent study by the FAA showed that pilots new to a particular GPS unit can spend as much as 45 percent of their time adding data, amending data, or otherwise staring at the GPS display and/or moving map. Who's flying the airplane while you stare at that little plane crawling across the moving map toward the wrong place?

Let's hope that you have an autopilot on board your trainer that can hold the plane steady while you work out the problem. But what are you going to do the day you rent a plane you've never used, and there's no working autopilot? Oh, and just to toughen the challenge more, your destination is only 100 feet above GPS minimums with no ILS.

You decide you're up for any challenge, so you hop in and take off. The turbulence outside and an annoying passenger inside combine to wear on you. To make matters worse, ATC has the audacity to change the approach in use after you've just

---

Shortcuts That Don't Shortchange-4

An excellent way to practice VNAV is on GPS VFR cross-country flights. Let's say you're cruising at 6,500 feet, and you want to end up five miles from your destination at traffic pattern altitude plus 1,000 feet. What rate of descent do you need to maintain if you begin descent from your present position? Then work the problem the other way: When do you begin descent if you want to maintain a 500 feet-per-minute rate of descent?

set up for the approach you planned originally. That's when it happens—you start loading the new approach and you hit the wrong button.

First of all, calm down and FLY THE AIRPLANE. Most veterans have heard "aviate, navigate, and communicate" as the pilot's workload triage mantra. Believe in it! It's always true.

The same applies here. Take a deep breath, go back into your instrument scan for a few seconds to make sure you're still in charge of the plane. Don't forget to keep the vacuum gauge in your scan, especially if you're on autopilot. Next, go to your favorite nav page on the GPS unit. What does it show as the next waypoint in your trip? If the wrong button didn't change the next waypoint, you are still working the previous plan.

Pick up your instrument scan again for another 10 seconds of sweep, then go to the airport page containing the newly cleared approach procedure. Selecting it and its initial approach fix will prompt you to confirm that you want to change approaches. Go ahead and change to your amended clearance, and be sure to verify that the next fix you're heading for still fits into the plans. If it doesn't, you should highlight the next cleared waypoint in the flight plan sequence and go ➡ (assuming that was your cleared route).

In the preceding scenario, if you did accidentally change the next waypoint to something different, verify with ATC what your next waypoint is and find it on the active flight plan page. Highlight it with the cursor and select ➡ . Verify again on the nav page for your flight that you have properly directed the unit to that waypoint.

Again, there's no more dangerous situation you're liable to be in than if you fixate on the GPS unit and let your instrument scan go. Keep a "fly it first" mentality and remember that you can always request ATC to give you a heading to fly while you work something out on the box. If you have an autopilot, use it. If you have a copilot, use him or her. If you're alone or with a nonpilot, trim the airplane well and step up the pace of your scanning.

Once the *aviate* part of the trio is taken care of, navigate with ATC's help. Use VOR as a backup to keep you in a situationally aware state. The VOR system is—and will be—an integral part of flight planning for years to come still. Since one VOR generally looks and operates like another VOR, the differences among them are less of a problem than with GPS units. Due to the variety of displays on GPS units sold today, there will always be a longer familiarization period than is necessary for VORs. When you've seen one VOR, you've seen them all (unless it's an HSI; but then they, too, all work similarly).

If you do switch to the VOR system from the GPS system, don't forget the "gotcha" button. As stated elsewhere in the book, this is the button that changes the input to your nav display from GPS to NAV. It may be called the GPS/NAV selector switch, and if there's room on your descent or approach checklist, write it in on your do-it-yourself checklist.

FAR Part 135 pilots who want to make this checklist change permanent would have to be approved by your company's FAA principle operations inspector, so don't just write it in without guidance from your chief pilot.

Also, STARs are based on VORs and will be for some time in the future. After all, a lot of planning went in to the locations of the VORs for various reasons. These systems will be scaled down or reduced in number as time goes by, but they will definitely not be eliminated completely in the near future. This is also true for VOR approaches. Due to the increased dangers of living in today's world, the VOR system is a comfortable safety net to have under us if something goes awry with the satellite system, if only for a brief period.

Use this to your advantage to find a VOR in front of or behind you that defines the segment of the STAR or approach you're on and duplicate your navigation. Then, if you hit a wrong button, your situational awareness doesn't suffer.

A lot of pilots would rather sit back and eat peanuts than go to this trouble on VFR trips. But if you make a practice of duplicating GPS and VOR usage—especially when you're not familiar with the GPS equipment—it will become second nature to switch to VOR navigation on those trips when the wonderful world of GPS is acting cantankerous.

In summary, remember the following:

1. *Fly the plane.* Make use of the autopilot or rated and current copilot. Be very aware of the amount of time you spend away from your instrument scan.
2. After verifying that all is well with your control of the airplane, go to a nav page that displays your next programmed waypoint to verify that it wasn't changed by the errant button push.
3. Use ATC to confirm your next point or to give you a vector to that point. Good cockpit resource management (CRM) often extends beyond the cockpit.
4. Use VOR as a backup system to enhance your situational awareness. That's good CRM, too.
5. If the error happened while changing approaches or STARs in the active flight plan, go to your unit's airport function and choose either the approach selection page or the arrival selection page, as applies, and make the correction you need.
6. Selecting a new arrival or approach will cause a message to pop up verifying that you do wish to update the flight plan to a new procedure. Choose YES if you're satisfied with your new setup.
7. If the error occurred in any other flying situation, simply go to a nav page you are comfortable with for verification that you're navigating according to your ATC clearance.

The number of button-pushing errors you will have on your flights can be reduced dramatically with more experience using the unit. Practicing with the GPS unit on the ground, either with a docking station at the FBO or on the ramp with a power cart hooked to the airplane as we've discussed previously will reduce your in-flight errors significantly.

Black Box Practice-2

1. Turn on your GPS. Check start-up screens for correct information. Enter correct baro and fuel on board.
2. Check to make sure that your nearest airport preferences are what you want.
3. Create a do-it-yourself checklist for creating, storing, and retrieving a flight plan.
4. Enter a flight plan with three or more waypoints using your checklist, store it, and then retrieve it.

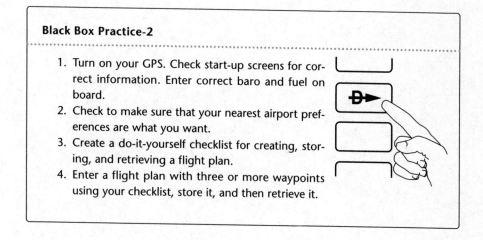

5

How to Plan a GPS Flight

It's okay to take a walk or go for a spin in your car without a plan, even on a rainy day. We do it all the time and don't give it a thought. But this obviously won't work on an IFR flight. Not only must you have a scheme for how to get up, how to get to a destination, and how to get down, but you should have a few contingency plans as well. The same is true for a VFR cross-country flight.

With good planning, the use of GPS on an IFR cross-country flight can, believe it or not, be much simpler in the long run than going back and forth between a panel display and detailed paper navigation logs, en route charts, and departure and approach plates. All it takes is some practice and some basic planning. Your GPS unit will take it from there. In GPS-land, your paper charts and detailed nav logs will likely stay wherever you put them when you organize your papers for the flight. Welcome to the *almost* paperless cockpit!

GPS Planning with DUATS

There is one piece of paper you must have, however, and that is something to carry into the cockpit that has the information to punch in to set up your GPS cross-country. The best way to come up with this crucial piece of paper, we believe, is to plan and print out your flight using the free flight planning computer program provided by the FAA's DUATS (direct user access terminal system). DUATS is available free of charge through two computer weather briefing and flight planning services: CSC DUATS (formerly DynCorp DUATS) and DTC DUAT.

If you are new to DUATS, you will find it helpful to walk through DUATS registration and flight planning with an experienced user working with you for the first couple of sessions. Some of the DUATS moves can be a little tricky at first.

If you don't have access to a computer, don't worry. Just plan a point-by-point VOR/airways route as you normally do. But file it with a "G" suffix and use GPS for navigation and VORs as backup. A little farther on we will show you how this works out. See Fig. 5.7.

Accessing DUATS

CSC DUATS and DTC DUAT are free computer services for all pilots with current medical certificates, plus others in a few categories that do not require medicals. To use DUATS you must first register with one or both services that provide it.

See Chap. 12, Reference Section (pages 143–144), for details on access numbers, how to register, services available through the two DUATS providers, and other details.

The DUATS FLIGHT PLANNER saves you time and effort by automatically making many essential calculations for you. You enter departure time, cruising altitude, aircraft information, and your routing preferences (select DIRECT ROUTING FOR RNAV to obtain a GPS great circle routing). You get back a flight log complete with distances, times, headings, and fuel burns all corrected for the current winds aloft at the altitude you want.

If you select DIRECT ROUTING FOR RNAV, DUATS will calculate the best-curved great circle route for you. This is all but impossible for most IFR pilots without special equipment.

DUATS will also automatically provide great circle distances over this curved route. Times en route for each segment of the great circle are automatically calculated on the basis of the winds aloft being reported for your planned altitude. For a short flight, the time saved is not all that much. But over a long haul the savings can be significant. See Fig. 3.2 again for the depiction of a KJFK–KLAX direct flight as computed and drawn by the CSC DUATS Cirrus program.

The DUATS FLIGHT PLANNER automatically computes the course changes needed to fly great circle routes. The planner also automatically corrects headings for winds aloft, as well as for compass variation. The headings and heading changes resulting from all three corrections—great circle, winds, and variation—are far more reliable and much more accurate than most human calculations. Why use anything else?

Required VOR Backup for GPS

To be even more helpful, the DUATS FLIGHT PLANNER also provides backup VOR stations for GPS flights. This is not just nice information to have in case of a GPS outage, it is *required* by the FAA. The reason is that GPS is still not certified as *primary* means of air navigation except in oceanic airspace and certain remote areas. Here is how the FAA puts it in AIM: "Properly certified GPS equipment may be used as a supplemental means of IFR navigation for domestic en route, terminal operations, and certain instrument approach procedures (IAPs)." (See Chap. 12, Reference Section, pages 161, for full text of statement.) Note that GPS is still referred to as a *supplemental*, not *primary*, system for IFR navigation.

In the real world of IFR, then, you must be prepared to switch from GPS to VOR navigation at any time. And buried in the fine print is yet another requirement: "Any required alternate airport must have an approved instrument approach procedure other than GPS that is anticipated to be operational and available at the estimated time of arrival, and which the aircraft is equipped to fly." (See "AIM Excerpts Relating to GPS" in Chap. 12, Reference Section, page 174.) You must have a means of getting down safely at your alternate without GPS if you have a RAIM problem throughout your geographical area or a GPS equipment outage of any kind.

But note: WAAS-enabled units have the advantage of allowing the pilot to file and use a GPS alternate since WAAS has its own system of integrity monitoring and does not require RAIM.

Sample GPS Flight Plan

Let's set up a real-world GPS IFR flight and see how easy it is to work out the details of a great circle flight using the DUATS format provided by CSC DUATS's Cirrus program.

As our real-world example, we have chosen a representative IFR flight—not too hard, not too easy—from Douglas International in Charlotte, NC (KCLT), to Dekalb-Peachtree (KPDK), a major satellite airport near The William B. Hartsfield International Airport (KATL) in Atlanta, Ga. All you have to do is access DUATS's FLIGHT PLANNER option and fill in the blanks, including those for your aircraft performance and personal profiles. (If you fill out the latter two items when you first begin using DUATS it will eliminate having to re-enter the same information each time you work with the FLIGHT PLANNER. Performance numbers and personal information can easily be updated anytime they change.)

Now look at Fig. 5.1 (next page) to see the completed FLIGHT PLANNER prepared by DUATS for this GPS-based flight in a Cessna 172 on a typical day in the Southeast. Check it over, change it if necessary, and then print it out. This is the piece of paper that will go with you into the cockpit with the information you need to punch in for your GPS flight—except for one very important item: DUATS *will not* determine an alternate for you. We really don't want a computer choosing our alternate for us! We'll cover alternates a little further on in this chapter.

Note how items 1 to 5 in the left-hand columns of Fig. 5.1 have been listed. The airport identifiers for Charlotte and Dekalb-Peachtree now have "K" in their codes. This is the ICAO (International Civil Aviation Organization) prefix for American airports in the contiguous United States. ICAO identifiers are now used universally in GPS databases. Thus Charlotte becomes KCLT and Dekalb-Peachtree becomes KPDK.

```
Cruise altitude: 60
  Flight level 60 accepted as 6000 feet
                    DYNCORP DUATS FLIGHT PLAN
From: KCLT -- Charlotte NC (Charlotte/Douglas Intl)
To:   KPDK -- Atlanta GA (Dekalb-Peachtree)
Alt.: 6,000 ft.
Time: Mon Apr 21 23:00 (UTC)
Routing options selected:  Great circle RNAV.
Flight plan route:
  SPA150011 ELW330004 AHN330018
Flight totals: fuel: 15 gallons, time: 2:28, distance 184.4 nm.
     Ident  Type/Morse Code   |                        | Fuel
     Name or Fix/radial/dist  |                        | Time
     Latitude Longitude Alt.  | Route   Mag  KTS  Fuel | Dist
  ---+--------+---------+-----|  Winds  Crs  TAS  Time |------
  1. KCLT   Apt.             |  Temp   Hdg   GS  Dist |  0.0
     Charlotte NC (Charlotte |--------+----+---+------|  0:00
     35:12:50  80:56:35    7 | Direct            4.1 |  184
  ---+--------+---------+-----| 238/22  252   95  0:39 |------
  2. RNAV   115.7/150.0/010.8 | +16C    251   72   47 |  4.1
     SPA     ...   .--.   .-  |--------+----+---+------|  0:39
--More--
     34:52:52  81:48:39   60 | Direct            4.6 |  137
  ---+--------+---------+-----| 260/28  247  100  0:46 |------
  3. RNAV   108.6/330.0/004.3 | +10C    251   73   55 |  8.7
     ELW     .  .-..   .--    |--------+----+---+------|  1:25
     34:28:47  82:49:37   60 | Direct            2.9 |   82
  ---+--------+---------+-----| 267/26  246  100  0:30 |------
  4. RNAV   109.6/330.0/017.9 | +10C    252   75   38 | 11.6
     AHN     .-   ....   -.   |--------+----+---+------|  1:55
     34:12:19  83:30:17   60 | Direct            3.2 |   44
  ---+--------+---------+-----| 246/21  246  100  0:33 |------
  5. KPDK   Apt.             | +15C    246   79   44 | 14.8
     Atlanta GA (Dekalb-Peac |--------+----+---+------|  2:28
     33:52:32  84:18:07   10 |                      |    0
  ---+--------+---------+-----|                      |------
```

NOTE: fuel calculations do not include required reserves.
Flight totals: fuel: 15 gallons, time: 2:28, distance 184.4 nm.
Average groundspeed 75 knots.

Recalculate plan at a different cruise altitude [Y/N]? [N]

Figure 5.1 CSC DUATS flight plan/nav log for hypothetical trip from Charlotte Douglas, N.C., to Dekalb-Peachtree, GA. Note how VOR backups are listed for this great circle route.

Items 2, 3, and 4 are the VORs that serve as the non–GPS backups along the great circle route from KCLT to KPDK. These three VORs also have another important GPS function. For precise GPS navigation along a great circle, you must also have a set of *waypoints* (or fixes) to make sure you are not drifting off your flight planned route.

NOS or Jeppesen Charts?

Both families of instrument charts are equally good and authoritative, and for the experienced instrument pilot, it all boils down to a matter of personal preference. However, for training purposes we prefer the government charts. For one thing, the charts illustrated in the instrument written test are chosen from NOS publications. Also, the charts used as examples in the FAA's *Instrument Flying Handbook* and AIM are NOS. We find it less confusing to use only one type of chart throughout all phases of instrument training.

Some waypoints will be facilities directly on a great circle track that you will fly over. But in the KCLT–KPDK example you do not fly *over* the VORs designated as a waypoint. You fly *past* them on the great circle route. When the waypoint is 90° off your wing, you should be on the radial indicated at the distance shown. These radial and distance offsets are given right after the name of the VOR. For example, the SPA (Spartanburg) waypoint is 150.0/010.8. It is located on the 150.0° radial from the station at 10.8 miles.

Now look again at the DUATS printout for your flight, and locate the FLIGHT PLAN ROUTE at the beginning of the printout (just above FLIGHT TOTALS). The flight plan route reads SPA 150011 ELW 330004 AHN 330018. This is what you will enter for ROUTE OF FLIGHT on the flight plan you file with ATC. This specifies the waypoints DUATS has set up along your great circle route, rounded off to the nearest whole numbers. Note that frequencies are listed for the VORs but no TO/FROM numbers are given. Since this is a GPS flight, you will proceed direct TO each waypoint without tracking outbound from the VOR.

Mark these waypoints on the Enroute Low Altitude chart L-20. (See Fig. 5.2, next page) and connect them. As you actually fly this trip, GPS will continuously indicate small heading adjustments to make a smooth great circle segment.

The right-hand set of columns shows the GPS numbers for this flight in the Cessna 172. All these numbers are based on your planned great circle track with winds aloft and compass variation factored in. Distances, times, headings, and fuel burns are automatically calculated for winds aloft. All distances are based on great circle segments. The time saved by flying a great circle is not very much on this flight, about 6 nautical miles. The time saved would become more of a factor on a longer flight such as KJFK–KLAX, even more on a trans-Pacific flight. The last time listed at the bottom of the far right column is cumulative and gives you the all-important total estimated time en route (ETE) for the flight plan you will file with ATC. Small heading changes periodically keep you on the great circle track, as well as correcting for winds.

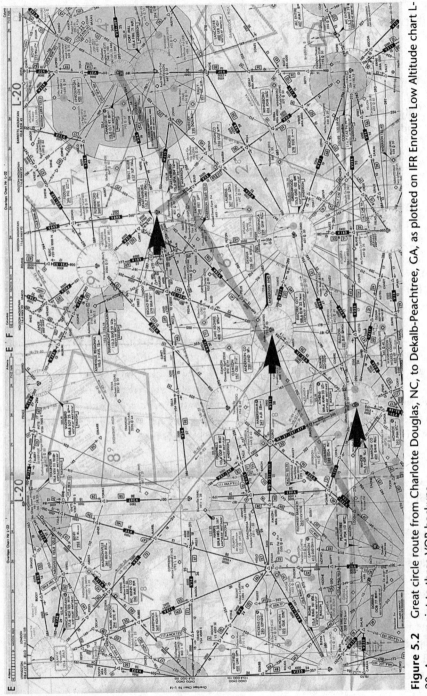
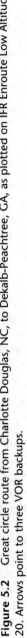

Figure 5.2 Great circle route from Charlotte Douglas, NC, to Dekalb-Peachtree, GA, as plotted on IFR Enroute Low Altitude chart L-20. Arrows point to three VOR backups.

Rehearsing the Route

While you have the L-20 chart handy, mentally trace the route worked out by DUATS. Note *all* the reporting points (triangles) and intersections along the way. Experience indicates that unless you do this, ATC may amend your clearance to go some place you never heard of before, creating world-class confusion in a single-pilot, single-engine situation, especially at night.

The same holds true for noting all the VORs, NDBs, and airports along the way. You never know when ATC will amend your clearance to move you along more expeditiously, hold you somewhere because of a delay, or reroute you because of a problem. As you will see in Chap. 6, these matters are handled easily on a GPS flight. But you still need to understand what ATC wants you to do before you start punching the buttons.

Departure Procedures (DPs)

Your GPS flight planning should always include a study of all the departure procedures (DPs) available at your departure airport. Why is this step necessary, you ask, when departure control is going to give you radar vectors to get you on your way?

There are two reasons. The first is to avoid nasty surprises. There is an IFR corollary to Murphy's law that says: "If you don't know where something is, you are almost certain to get sent there!"

Second, in the rare event of a two-way communications failure, you must be able to proceed according to your latest clearance without any more radar vectors. If your last clearance is a published DP, for example, you will need to be familiar with its checkpoints and crossing altitudes without too much fumbling around with charts and booklets.

DP charts are found in NOS approach plate booklets immediately after approach plates and airport diagrams (if any). An alphabetical list of DPs also appears in the index at the front of each set of NOS approach plates.

At the present time, the majority of DPs are overlays of VOR-based procedures. As time goes by, you will see more DPs based solely on GPS navigation. Such a DP is the CINGA departure for Juneau, Alaska (KJNU). Note the (RNAV) just after the title.

See Fig. 5.3 (next page) for the approach plate index page with the DPs listed for Charlotte International. There are three DPs for Charlotte, and it won't take you long to realize there is really only one DP for all runways for propeller aircraft. This is the HUGO FIVE DEPARTURE (Figs. 5.4*a* and 5.4*b*, pages 45, 46). Study the transitions and make sure you understand which one you are most likely to get. How about HUG5.DEBIE? But take a look at the other two departures just in case ATC pulls a surprise and gives you something else.

INDEX
03359

INDEX OF TERMINAL CHARTS AND MINIMUMS

NAME	PROC	SECT PG

CHARLESTON, SC
CHARLESTON AFB/INTL(CHS)
TAKE-OFF MINIMUMS C
ALTERNATE MINIMUMS .... E
RADAR MINIMUMS .... N
IAPS ..... ILS RWY 15 69
ILS RWY 33 70
ILS RWY 15(CAT II) 71
RNAV (GPS) RWY 3 72
RNAV (GPS) RWY 15 73
RNAV (GPS) RWY 21 74
RNAV (GPS) RWY 33 75
VOR/DME OR TACAN RWY 3 76
VOR/DME OR TACAN RWY 15 77
VOR/DME OR TACAN RWY 21 78
VOR/DME OR TACAN RWY 33 79
NDB RWY 15 80
AIRPORT DIAGRAM 81

CHARLESTON EXECUTIVE(JZI)
TAKE-OFF MINIMUMS C
IAPS ..... VOR/DME RNAV RWY 9 82
VOR OR GPS-A 83
NDB RWY 9 84
GPS RWY 9 85

CHARLOTTE, NC
CHARLOTTE/DOUGLAS INTL(CLT)
TAKE-OFF MINIMUMS C
ALTERNATE MINIMUMS .... E
STARS .. CHESTERFIELD TWO P6
MAJIC NINE P8
SHINE FIVE P11
UNARM ONE P14
IAPS ..... ILS RWY 5 86
ILS RWY 18L 87
ILS RWY 18R 88
ILS RWY 23 89
ILS RWY 36L 90
ILS RWY 36R 91
ILS RWY 36L(CAT II) 92
ILS RWY 36R(CAT II) 93
ILS RWY 36L(CAT III) 94
ILS RWY 36R(CAT III) 95
RNAV (GPS) RWY 5 96
RNAV (GPS) RWY 18L 97
RNAV (GPS) RWY 18R 98
RNAV (GPS) RWY 23 99
RNAV (GPS) RWY 36L 100
RNAV (GPS) RWY 36R 101
VOR RWY 36L 102
NDB RWY 5 103
AIRPORT DIAGRAM 104
DPS ..... HORNET ONE 105
HUGO FIVE 106
PANTHER FIVE 108

CHERAW, SC
CHERAW MUNI/LYNCH BELLINGER FIELD(47J)
TAKE-OFF MINIMUMS C
IAPS ..... VOR/DME OR GPS RWY 7 110
NDB RWY 25 111
GPS RWY 25 112

CHERRY POINT MCAS (CUNNINGHAM FIELD)(KNKT)
CHERRY POINT, NC
TAKE-OFF MINIMUMS C
RADAR MINIMUMS .... N
IAPS ..... ILS RWY 23R 113
TACAN RWY 32L 114
AIRPORT DIAGRAM 115

Figure 5.3 Index page B3 of NOS Approach Chart booklet lists many GPS approaches for Charlotte Douglas, NC.

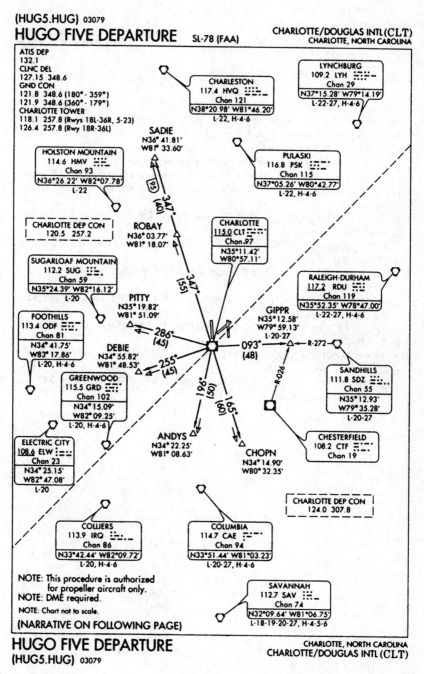

ATIS DEP
132.1
CLNC DEL
127.15 348.6
GND CON
121.8 348.6 (180° - 359°)
121.9 348.6 (360° - 179°)
CHARLOTTE TOWER
118.1 257.8 (Rwys 18L-36R, 5-23)
126.4 257.8 (Rwy 18R-36L)

LYNCHBURG
109.2 LYH
Chan 29
N37°15.28' W79°14.19'
L-22-27, H-4-6

CHARLESTON
117.4 HVQ
Chan 121
N38°20.98' W81°46.20'
L-22, H-4-6

SADIE
N36° 41.81'
W81° 33.60'

HOLSTON MOUNTAIN
114.6 HMV
Chan 93
N36°26.22' W82°07.78'
L-22

PULASKI
116.8 PSK
Chan 115
N37°05.26' W80°42.77'
L-22, H-4-6

CHARLOTTE DEP CON
120.5 257.2

ROBAY
N36° 03.77'
W81° 18.07'

CHARLOTTE
115.0 CLT
Chan 97
N35°11.42'
W80°57.11'

SUGARLOAF MOUNTAIN
112.2 SUG
Chan 59
N35°24.39' W82°16.12'
L-20

PITTY
N35° 19.82'
W81° 51.09'

RALEIGH-DURHAM
117.2 RDU
Chan 119
N35°52.35' W78°47.00'
L-22-27, H-4-6

FOOTHILLS
113.4 ODF
Chan 81
N34° 41.75'
W83° 17.86'
L-20, H-4-6

GIPPR
N35° 12.58'
W79° 59.13'
L-20-27

286°
(45)

093°
(48)

R-272

R-026

DEBIE
N34° 55.82'
W81° 48.53'

255°
(45)

SANDHILLS
111.8 SDZ
Chan 55
N35° 12.93'
W79° 35.28'
L-20-27

GREENWOOD
115.5 GRD
Chan 102
N34° 15.09'
W82° 09.25'
L-20, H-4-6

196°
(50)

165°
(60)

ELECTRIC CITY
108.6 ELW
Chan 23
N34° 25.15'
W82° 47.08'
L-20

ANDYS
N34° 22.25'
W81° 08.63'

CHOPN
N34° 14.90'
W80° 32.35'

CHESTERFIELD
108.2 CTF
Chan 19

CHARLOTTE DEP CON
124.0 307.8

COLLIERS
113.9 IRQ
Chan 86
N33°42.44' W82°09.72'
L-20, H-4-6

COLUMBIA
114.7 CAE
Chan 94
N33°51.44' W81°03.23'
L-20-27, H-4-6

NOTE: This procedure is authorized
for propeller aircraft only.
NOTE: DME required.
NOTE: Chart not to scale.

(NARRATIVE ON FOLLOWING PAGE)

SAVANNAH
112.7 SAV
Chan 74
N32°09.64' W81°06.75'
L-18-19-20-27, H-4-5-6

Figure 5.4a   Plan view of HUGO FIVE departure procedures (DPs) at Charlotte Douglas, NC.

(HUG5.HUG) 02276
HUGO FIVE DEPARTURE SL-78 (FAA)
CHARLOTTE/DOUGLAS INTL (CLT)
CHARLOTTE, NORTH CAROLINA

▼
DEPARTURE ROUTE DESCRIPTION

TAKE-OFF ALL PROPELLER AIRCRAFT ALL RUNWAYS: Fly assigned heading and maintain 4000 feet for vector to intercept assigned radial associated with issued transition. Proceed via the depicted radial to the transition fix, thence as filed. If no transition assigned, expect vector to appropriate fix.

Expect filed altitude/flight level 10 minutes after departure.

NOTE: Runway 36R and Runway 5 departures are required to climb at 240 feet per nautical mile to 1900 feet. If unable to accept this climb rate advise ATC on initial contact.

ANDYS TRANSITION (HUG5.ANDYS): From over CLT VOR/DME via CLT R-196 to ANDYS INT. Thence as filed.

CHOPN TRANSITION (HUG5.CHOPN): From over CLT VOR/DME via CLT R-165 to CHOPN INT. Thence as filed.

DEBIE TRANSITION (HUG5.DEBIE): From over CLT VOR/DME via CLT R-255 to DEBIE INT. Thence as filed.

GIPPR TRANSITION (HUG5.GIPPR): From over CLT VOR/DME via CLT R-093 to GIPPR INT. Thence as filed.

PITTY TRANSITION (HUG5.PITTY): From over CLT VOR/DME via CLT R-286 to PITTY INT. Thence as filed.

ROBAY TRANSITION (HUG5.ROBAY): From over CLT VOR/DME via CLT R-347 to ROBAY INT. Thence as filed.

SADIE TRANSITION (HUG5.SADIE): From over CLT VOR/DME via CLT R-347 to SADIE INT. Thence as filed.

HUGO FIVE DEPARTURE
(HUG5.HUG) 02276
CHARLOTTE, NORTH CAROLINA
CHARLOTTE/DOUGLAS INTL(CLT)

Figure 5.4b HUGO FIVE departure descriptions provide actual clearance wording for departure procedure (DPs) at Charlotte Douglas, NC.

Standard Terminal Arrival Procedures (STARs)

Go through the same process for analyzing your destination STARs (Standard Terminal Arrivals). STARs are listed by airport in the index in the front of the NOS approach plates and are grouped together separately in the "P" section. Again, you will quickly see that there is one STAR for Dekalb-Peachtree airport that makes the most sense for traffic arriving from the east. And that is AWSON ONE ARRIVAL with a CERAY TRANSITION. See Figs. 5.5*a* and 5.5*b* (next two pages.).

Enter the STAR of your choice on your flight plan, and punch it in when you load your flight data into the GPS. But have a look at the other STARs available, just in case.

Anticipating the GPS Approach

The wind direction almost always determines the specific runway and instrument approach procedure (IAP) at the destination. For our hypothetical flight from KCLT to KPDK, the winds are southwesterly. So the logical landing runways at KPDK are 20L, 20R, or 27. (See the airport diagram, Fig. 5.6, page 50.)

Study the airport diagram first, just as you would for a VFR cross-country flight. Note the field elevation and the number of runways and their headings, lengths, and widths. Will you be able to use all runways? Do mountains or other obstructions affect an instrument approach? How high are the obstructions? Look over the taxiways from your landing runway to your destination on the field. Getting to your destination after landing at an unfamiliar field can sometimes be confusing—or embarrassing—especially at night!

Check the index page at the front of the NOS booklet and see what GPS instrument approaches are available for the westerly runways 20L, 20R, or 27. The obvious choice here is 20L. It is long and wide, and it has two GPS approaches and three non-GPS approaches available as backups.

There are also instrument approaches to the shorter westerly runway 27. So check these out as well. Remember the IFR corollary to Murphy's law: "If you don't know where something is, you're almost certain to get sent there!" And don't forget to fix the missed approach procedures (MAPs) in your mind.

Your selected approach, GPS RWY 20L, should be punched into the GPS right after the STAR. If you load an approach into the GPS at this point it will make the workload easier when things get really busy at the other end. You might even get lucky and end up with exactly what you planned for!

With your departure, en route, arrival, and approach planning completed, you now have *almost* enough information to complete your flight plan/nav log. One major item remains to be resolved, and that is your alternate airport. Now is the best time in your preparation for a complete weather briefing. You may obtain your brief-

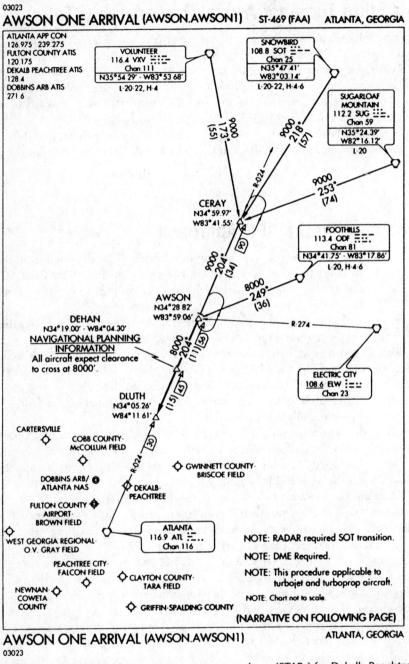

Figure 5.5a Plan view of AWSON ONE ARRIVAL procedures (STARs) for Dekalb-Peachtree, GA.

00111
AWSON ONE ARRIVAL (AWSON.AWSON1)   ST-469 (FAA)   ATLANTA, GEORGIA

ARRIVAL DESCRIPTION

<u>CERAY TRANSITION (CERAY.AWSON1)</u>: From over CERAY INT via ATL R-024 to AWSON INT. Thence. . . .

<u>FOOTHILLS TRANSITION (ODF.AWSON1)</u>: From over ODF VORTAC via ODF R-249 to AWSON INT. Thence. . . .

<u>SNOWBIRD TRANSITION (SOT.AWSON1)</u>: From over SOT VORTAC via SOT R-218 and ATL R-024 to AWSON INT. Thence. . . .

<u>SUGARLOAF MOUNTAIN TRANSITION (SUG.AWSON1)</u>: From over SUG VORTAC via SUG R-253 and ATL R-024 to AWSON INT. Thence. . . .

<u>VOLUNTEER TRANSITION (VXV.AWSON1)</u>: From over VXV VORTAC via VXV R-173 and ATL R-024 to AWSON INT. Thence. . . .

. . . .From over AWSON INT via ATL R-024 to DLUTH INT. Expect radar vectors to final approach course after DEHAN INT.

AWSON ONE ARRIVAL (AWSON.AWSON1)   ATLANTA, GEORGIA
00111

Figure 5.5b   AWSON ONE ARRIVAL descriptions provide actual clearance wording for Dekalb-Peachtree, GA, arrivals (STARs).

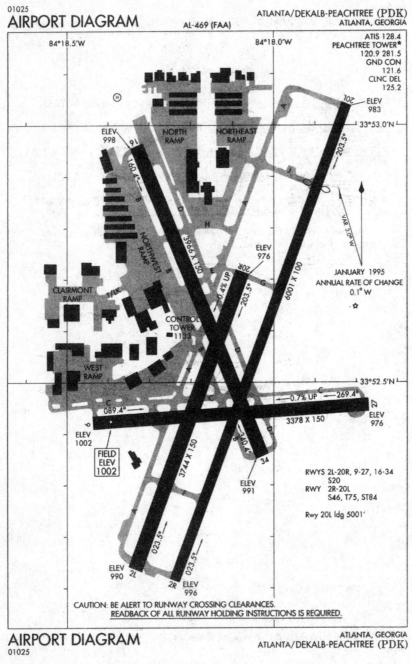

Figure 5.6 Dekalb-Peachtree, GA, airport diagram.

ing by selecting one of the DUATS weather options or by calling 1-800-WX-BRIEF.

Weather Factors

Can you make this flight safely and confidently? Will any of these five go/no-go weather factors affect the flight you have planned? If so, think seriously about changing your plans.

- Thunderstorms
- Turbulence
- Icing
- Fog
- Low ceiling and visibility at departure and/or destination airports

Begin your weather briefing by selecting STANDARD WX: ROUTE from the DUATS choices. This should provide more than enough information—including NOTAMS—for most IFR go/no-go decisions, as well as for choosing an alternate. Pick the handy PLAIN LANGUAGE option if the weather shorthand abbreviations drive you crazy. Go on to one of the other choices, such as STANDARD WX: AREA, if you need more information.

Choosing an Alternate

Do you need an alternate? A quick review:

If, for one hour before ETA through one hour after ETA, the ceiling and visibility are forecast to be 2,000 feet and 3 miles or less, according to FAR 91.169 (b), a pilot must select an alternate airport and include the selection on the flight plan. This is the easy-to-remember "one, two, three rule": *one* hour before and after, *two* thousand feet ceiling, and *three* miles visibility. As noted earlier, alternates must also have non-GPS approaches available.

DUATS will *not* pick an alternate, nor will it show you how to get there. So how do you go about this? First, study the weather compilations and see what might work for your specific situation. Obviously, you can't pick an alternate that is so far away that you'll run out of fuel before you get there. And there's not much point in picking an alternate where the weather is so bad it prevents a safe landing.

On these points the FARs are grounded in good common sense. First, FAR 91.167 says you must carry enough fuel on an IFR flight to:

- Complete the flight to the first airport of intended landing
- Fly from that airport to the alternate, if one is required
- Fly for 45 minutes at normal cruising speed after reaching the alternate airport (and taking into consideration ATC delays and the time it will take to make approaches at the destination and alternate)

You can list an airport as an alternate only if the ceiling and visibility forecast for the alternate at estimated time of arrival will be at or above 600 feet and 2 miles if the alternate has a precision approach, or 800 feet and 2 miles if it has only nonprecision approaches [FAR 91.169 (c)].

If you are WAAS equipped, you may use 600 feet and 2 miles when qualifying the weather if your alternate destination has either LPV or LNAV/VNAV minima, because they are considered precision approaches.

Watch out for a "catch-22" regarding these requirements. Some airports might not be authorized for use as alternates under *any* circumstances, while others might have *higher* minimum requirements than 600/2 and 800/2 because of local conditions such as hills, towers, or radio masts. How can you find this out?

Turn to the front of the NOS approach chart booklet and find the listing of "IFR Alternate Minimums." Airports with minimums that deviate from the standard 600/2 and 800/2 will be listed in this section. If your choice of an alternate is not listed, use 600/2 and 800/2, as discussed above.

Personal Minimums

Many competent instrument pilots set higher minimums for themselves than the published minimums. This is very smart! A brand-new instrument pilot might want to start out with a 1,000 foot ceiling and 3 miles visibility (VFR) minimums until gaining more experience and confidence. This can be lowered, depending on frequency of flights, until reaching the lowest minimums allowable.

If you don't fly very often and are barely meeting proficiency requirements, it might be prudent to raise personal limits to 500 and 2, for example. Many pilots make it a policy to fly instruments every three or four weeks. Many also sign up for instrument instruction every three months if they feel a little rusty.

Finally—and you won't find this in any FAA publications—where can you find reliable VFR conditions when the chips are down? Where can you go for a safe VFR landing if all electrical power is lost, and radios, transponder, GPS, electrically powered instruments, pitot heat, and lights all go dark? Where can you go if an instrument flight becomes truly horrendous?

Always learn where the nearest VFR conditions are so you can safely abort the IFR flight and land VFR. This sound advice comes from veteran instructor Henry Sollman, author of *Mastering Instrument Flying.* A weather briefer will provide this information when requested if it is not clear from the forecasts. If there is no VFR weather at an airport within range, don't go.

With access to Internet aviation weather services, such as that available through AOPA's Web site, you can use weather depiction charts for a graphic view of where you can find areas with ceilings greater than 3,000 feet and visibility greater than 5 statute miles.

"When in doubt, wait it out!"

Filing the GPS Flight Plan

Return now to the DUATS FLIGHT PLANNER and note the line below the flight plan that says USE THIS INFORMATION TO FILE A FLIGHT PLAN [Y/N]. Enter Y for yes, and respond to the prompts that follow. Supply any additional information, including your choice of an alternate, if required. Then file your flight plan, and it will be transmitted immediately to the nearest regional Air Route Traffic Control Center (ARTCC). Be sure to cancel this flight plan if was for practice only and you do not intend to take off.

Print out the whole exercise, cut off all but the essential data you need for the flight, and take these numbers, including the flight plan you have filed, into the cockpit.

The good news is that it takes much less time to plan and file an IFR flight with DUATS than it does to talk about it! And here's more good news—most designated examiners will accept DUATS flight plan/nav log printouts if you get their OK ahead of time when setting up a check ride.

Planning GPS Flights without DUATS

If you don't have a computer handy, you won't be able to access DUATS. But planning GPS IFR flights without DUATS simply returns you to familiar territory—you plan your trip the same way you have been doing it all along, using airways and VORs.

Turn again to the L-20 IFR Enroute Low Altitude chart, and plot the shortest route from KCLT to KPDK using Victor airways and VORs as much as possible. How about:

V54 SPA V415 ODF V222 WOMAC

See Fig. 5.7 (next page). This route is approximately 6 miles longer than the GPS great circle route worked out earlier, and you will have to create your own flight

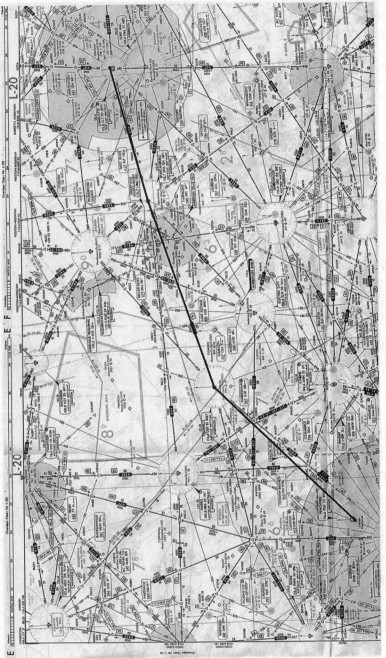

Figure 5.7 If you do not have access to a computer, plan and file a point-to-point GPS route.

plan/nav log to carry into the cockpit. One advantage in using this method is that your non-GPS backups are the same as your GPS waypoints as you proceed from point to point.

Do you have a flight plan/nav log form that you already feel comfortable using? If not, there are several acceptable versions in preprinted pads available through Sporty's and similar outlets. *Mastering Instrument Flying* also has an excellent form developed by author Henry Sollman that he invites you to copy.

Black Box Practice-3

1. Turn on your GPS. Check start-up screens for correct information. Check preference pages to make sure the default choices are what you want.
2. Punch in the numbers for the DUATS version of the KCLT–KPDK flight on your GPS unit. Repeat until you have the process down to less than five minutes without fixating on the display.
3. Punch in the numbers for the non-DUATS version of the KCLT–KPDK flight on your GPS unit. Repeat until you have the process down to less than five minutes without fixating on the display.

6

DEPARTURE, EN ROUTE, AND ARRIVAL

Our discussion of en route GPS flying begins at Douglas International Airport in Charlotte, NC. This particular day is 1,500 feet overcast with visibility about 4 miles in light rain and mist. There is even some talk of the embedded thunderstorms along the route to our destination, Dekalb-Peachtree Airport in Atlanta, GA.

You can use GPS in two ways as you travel across the country. If VFR conditions prevail, you can use a yoke-mounted or a handheld GPS unit. A panel-mounted unit is better, of course, since you can use it for both VFR and IFR. You can also couple panel-mounted GPS to a CDI or an HSI. An autopilot can then be slaved to the CDI or HSI.

But using a yoke-mounted or handheld unit is illegal for IFR. So your GPS must be panel-mounted to cope with the IFR conditions you are likely to encounter on a day like this. There is another aviation variation on Murphy's law operating here: "If you're not prepared for IFR, you are sure to get it!"

As you approach the airplane, you should have your mind focused on what things make an airplane safe for flight. You have worked out your flight plan/nav log and filed the flight plan with ATC before leaving the FBO briefing room. So, now is the time to make sure that the pitot heat works, the fuel tanks are full and free of water, and everything else is ready to go. In a few minutes you'll have the chance to enter your navigation data into the GPS, but for now, take one step at a time. Think GPS IFR as you go down your checklist. Did you check the GPS antenna, for example?

And, oh, by the way, do you know where your AFM (aircraft flight manual) is? Not the handy little "owner's manual"—but the official FAA-approved AFM manual for your airplane? Your friendly FAA inspector will insist on seeing this when he or she shows up for a surprise "ramp check."

If the AFM isn't handy in the cockpit, you've got a problem. But let's say you find it. You're still not off the hook. Does the AFM include a supplement covering the GPS installed in the plane? And does that GPS supplement make reference to a GPS quick reference or handbook? If a supplement requires items such as these, they had better be there within arm's reach.

If you do not have these legally required documents "immediately available to the flight crew" in your plane, you may be racking up some violations! It pays to be thoroughly familiar with your airplane's AFM and its supplements.

GPS Cockpit Checks

After completing the walk-around inspection of your Cessna 172, you take a comfortable seat behind the left yoke. Should you start up the GPS and then enter the flight plan data, or should you start up the airplane engine and then enter the data? Well, that old battery may or may not have enough juice left if you ran the flaps down and then up, checked the lights and pitot heat, listened to ATIS, and copied your clearance. This may be a poor time to have to get a jump-start—or to lay over while the battery is recharged, maybe even replaced. It's your call!

Entering GPS data begins with turning on the master switch, the avionics master switch, and then the GPS unit. As the GPS powers up, it will run through an important series of self-tests. The self-test function is the same from one manufacturer to another. But the displays and the numbers differ substantially among manufacturers. For discussion purposes, let's use the Bendix/King KLN 94 as our start-up example.

After switching the GPS unit on, a screen will appear with a code stating the revision of the GPS software. Not to be confused with the database software, this is the ORS (operational revision status) code, which should match the code of the handbook or quick reference you are required to carry.

When the KLN 94 internal self-tests are complete, a screen pops up on the GPS display showing a CDI bar with an arrow pointing halfway between the center of the bar and the right edge of the bar (see Fig. 6.1). This is a correct indication.

Set the GPS/NAV selector on the instrument panel to GPS. This will feed GPS navigation data to the CDI or HSI from the GPS unit. Look at that CDI or HSI. If it is coupled correctly to the GPS unit, it will also show an arrow between the center of the CDI bar and the right edge of the CDI bar. On an HSI, it will indicate the same provided that you have spun the needle around to point toward the nose of the airplane (straight up as you look at the instrument).

Leave the GPS/NAV selector in the GPS position (either the button or a nearby annunciator panel will display the letters GPS) if you are planning to make the departure using GPS.

Going back to the KLN 94 display, the next thing you'll notice is that the mileage at top left will read 34.5 nm if the display is functioning properly. The RMI field will read 130°. If you have a coupled RMI unit in the airplane, it will also read 130°.

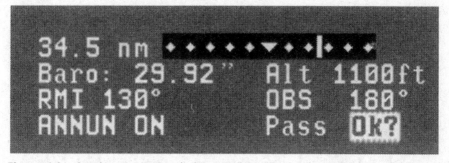

Figure 6.1 Starting screen, Bendix/King KLN 94. (Copyright © 2004 Honeywell International, Inc. All Rights Reserved.)

While this screen is still being displayed, reset the baro to match the latest altimeter setting you have copied from ATIS. It's a good habit always to reset the baro every time you get a new altimeter setting from ATC or ATIS.

The display on this, and subsequent start-up screens, will ask you to OK? or ACKNOWLEDGE? the data if they are correct. If the data are correct, press ENTER to move on.

On all units, regardless of manufacturer or model, your GPS will cycle to a display that verifies the position of the airplane by its latitude and longitude. This should match the latitude and longitude for the airport where the plane is located when the GPS is turned on (see Fig. 6.2, next page).

You'll find the latitude and longitude of your start-up airport on the approach plates, the en route charts, or in the *Airport/Facilities Directory* (AFD). Here you are checking to make sure the airplane really is where the GPS thinks it is. Acknowledge that the latitude and longitude are correct and continue to the next screen. If the position is not correct, reset the airplane's position according to the steps in your operating manual. (But ignore small differences. You may be parked some distance from the point on the field where the airport location is benchmarked for chart purposes.)

One of the most important things the self-tests will show is database currency. Would you take off IFR with an out-of-date database? Don't even think about it! Do not make a GPS flight if the database self-test display shows an expired database (see Fig. 6.3, next page).

At the completion of the self-tests, the display will cycle to either an airport information page or a flight plan page, depending on the unit you have. This is the starting point for entering your flight plan numbers.

But wait a minute! There are some other preflight items you must check on every GPS flight:

- You must make sure the fuel on board as shown on the GPS display matches the amount shown on your fuel gauges.

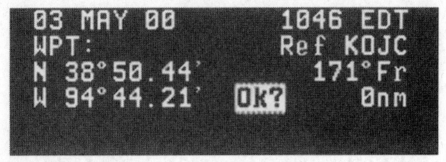

Figure 6.2 Latitude/longitude screen, Bendix/King KLN 94. (*Copyright © 2004 Honeywell International, Inc. All Rights Reserved.*)

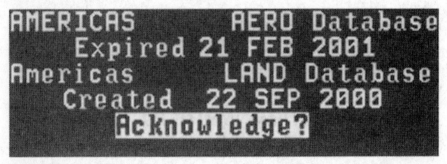

Figure 6.3 Database expiration screen, Bendix/King KLN 94. (*Copyright © 2004 Honeywell International, Inc. All Rights Reserved.*)

- The runway length and type of surface preferences for your NRST airport function must be set up the way you want them in case of an unexpected diversion.
- Your system may also require you to enable "turn anticipation" in order to use it. Now is the time to make sure this function is enabled.

You can get through the details of GPS self-tests and preflight items much more quickly—and with less chance of missing something important—if you make up a do-it-yourself checklist for these functions as described in Chap. 2.

Instructors Note: New GPS users—and some experienced ones as well—will often take the GPS self-tests and preflight checks too much for granted. This is not a good habit to get into. If you flip through these important items too fast, you can easily miss an equipment problem, an out-of-date database, an inaccurate fuel or baro setting, an inappropriate "nearest" default, or a disabled turn anticipation function in some systems. This is especially true in rental planes, and you don't want to take off depending on someone else's readings and defaults!

Help your students work out do-it-yourself checklists for self-tests, defaults, and other preflight items. Then use a challenge-and-reply method on every flight to verify all the checklist items. This is a really good habit for students to take with them as they work up the ladder into more complex airplanes and systems.

Now let's enter the numbers on that piece of paper you brought into the cockpit. Again, entering a flight plan is a function offered by all GPS systems. But the steps and displays can differ substantially. For consistency, we will continue with the Bendix/King KLN 94 as our example here. But no matter what system you use, you will find that a do-it-yourself checklist will help considerably in entering your flight plan data quickly and correctly.

Select the flight plan function (FPL in this case). Choose FPL 0 (the active flight plan), not a numbered flight plan stored from a previous flight. (See Fig. 6.4.) Enter the waypoints listed on the flight plan or nav log that you brought into the cockpit with you. List the waypoints sequentially to minimize confusion. Begin with the departure airport and end with the destination airport. Add your anticipated DP and STAR in their correct places in the sequence. Enter your anticipated approach procedure.

Equipment Note: The Bendix/King KLN 94 and Garmin 530 create and activate flight plans through the FPL 0 function. The Apollo GX60 creates and activates through the active page function. All three systems allow you to store, modify, and retrieve 20 or more flight plans.

A very useful feature of GPS is the ability to store flight plans in the GPS unit. For the pilot who goes to the same destination frequently, this will save a lot of time otherwise spent setting up the unit. To use this feature, go to the flight plan function and bring up the page where the numbered flight plans are stored. With the KLN 94, for example, turn the small knob clockwise from FPL 0 to FPL 1 through

Figure 6.4 Flight plan screen, Bendix/King KLN 94. *(Copyright © 2004 Honeywell International, Inc. All Rights Reserved.)*

FPL 25 to find a blank page. Enter your complete data on this blank page. When you finish entering the waypoints, the unit will prompt you to make this the active page. By answering yes, you not only store the numbered flight plan you just made, but it also becomes FPL 0—which is your active page for the flight.

In the real world of IFR, no matter how carefully you plan a flight, the universal experience is that something will change when ATC gets involved. ATC will *always* try to accommodate the flight you planned. But ATC must also maintain safe separation in traffic concentrations around extremely busy airports. The ATC controllers are responsible for maintaining an efficient traffic flow along published routes as well as when pilots elect to fly off airways on "GPS direct" flight plans. And they must reroute this traffic whenever severe weather conditions begin to slow things down.

So ATC has to cope with many variables you might not know about as its controllers try to fit your flight plan into the flow of traffic around your departure airport. Learn to expect something different from what you filed. It's the norm! Occasionally you'll get something that's even better than what you asked for!

It helps to plan for the most complex departure. This will probably be a published DP. Leaving Charlotte, for example, plan for the HUGO FIVE departure procedure with the appropriate transition to get you on course. See the two pages shown in Figs. 5.4*a* and 5.4*b*.

It is much better to add a complex departure procedure when you have time during data entry as opposed to later when you are busy trying to set up the plane for takeoff. If you have entered the most complex departure, then any change in departure is bound to be easier to handle. Leaving Charlotte, for example, departure control could simply clear you via radar vectors to your first en route waypoint to climb and maintain your requested altitude.

Copying Clearances

Contact clearance delivery for your IFR clearance after you have entered your flight plan data. If your plan is to proceed direct, be prepared for an amended clearance to fit you into the traffic flow more expeditiously. This is particularly true in the busy Northeast corridor on a weekday. On the trip we have planned, however, let's say we have a pretty normal clearance.

Remember that the purpose of this clearance is to get you airborne and to give you and ATC a plan in the event you lose all communications. If you want to go a more direct route once airborne, ask for it when you are comfortably on your way. ATC has become more used to seeing the "/GPS" equipment suffix these days and is more likely to give a pilot an amended clearance direct to the destination when the busiest time of the entire flight is over. So, accept the clearance given to you along with any changes ATC has made.

Shortcuts That Don't Shortchange-5

A favorite trick of experienced pilots when copying clearances is to spell out vertically the word

C

H

A

R

T

Beside the C, write in the words "as filed" and "radar vectors" (or AF and RV). When ATC says either "cleared as filed" or "radar vectors," you will only need to circle the appropriate words or letters. Next, beside the H, leave room to enter a heading—which may be as simple as the letters RH (for runway heading) or a number for a particular heading. A is for altitude, which comes next in the clearance sequence. For example, "two thousand, expect ten thousand ten minutes after departure" is written simply as 20 × 100 n10. Beside the letter R, write in the departure radio frequency followed by whichever departure control you will be talking to, for example, "120.5" and "Charlotte departure." If your unit has an integrated NAV/COM, your radio frequencies will be inserted automatically. Before you use them though, you should verify that the system put in the correct frequency sequence. Finally, T is for transponder. Simply write in the numerical code. The point in streamlining clearance copying is to help you with the readback of the clearance and catch any changes ATC gives you. This technique works just as well flying a Cessna as it does flying a Boeing 747.

Your next task is to prepare your cockpit for takeoff. Go straight down the clearance you just copied and make sure that every item has been set up correctly.

This brings up the subject of cockpit confusion.

Cockpit Confusion

Organize charts and flight plans or nav logs to avoid fumbling for a vital piece of information. A good device for holding charts and logs is a standard 8½- by 11-inch or 8- by 14-inch clipboard. Stick a hook piece of Velcro on the metal clip and loop pieces of Velcro on your favorite writing instrument so you will always be able to find it on a bumpy trip. A folding flight desk that rests on the lap or a kneeboard that straps to a thigh also work well.

Place the flight plan/nav log on top. Underneath this, place the en route chart folded to show the route, airways, and waypoints that might be in an amended clearance. Approach charts for the departure airport are placed under the en route chart in case you have to return for an instrument approach shortly after takeoff.

Next come approach charts for the destination airport with the expected approach on top. After these come all approaches for the alternate airport. You can save yourself some time and fumbling around in the cockpit if you have the flight plan/nav log and all your charts organized and clipped together before you leave the briefing room.

Finally, you should stash a sectional chart, folded to show your route of flight, so that you can get at it easily if you need to determine the OROCA (off-route obstacle clearance altitude) or to check the boundaries of some special use airspace (SUA). These suggestions add to your cockpit resource management (CRM). The ability to find things in a crunch by habitually putting them in the same predictable place is an invaluable way to be your own copilot.

If your GPS has been set up for the current flight plan and you are planning to use GPS as your main source of navigation information, verify that the GPS/NAV selector switch is set to the GPS position. If you're planning to use GPS as a backup to your NAV, check the selector switch in NAV position.

Next, if you have a heading bug, set it for the departure heading. Then, spin the needle of your CDI or HSI around to match the desired track (DTK) you expect to fly to your first waypoint.

The radios are next. Set up all of your communication and navigation frequencies in the correct sequence for the departure segment of your flight. Keep a pad of paper handy for tracking frequencies as you fly in the event you have to go back to a previous frequency for whatever the reason.

Before-Takeoff and Takeoff Checks

The run-up pad is a good place to take a serious look at the ammeter to make sure you are not draining power from the battery. Then run down the remaining items on the checklist. Just before adding power for takeoff do a mental STP check: strobes, transponder, pitot heat. Don't forget to write down your time of departure so that you can estimate your time of arrival at your destination and your fuel burn along the way.

Are you using a departure procedure (DP)? One of the capabilities of the GPS is to allow you to increase the sensitivity of the display when departing through congested airspace. By pushing the APPROACH ARM button you can increase the sensitivity from 5 miles down to 1 mile. Since a terminal area, such as Douglas Airport in Charlotte, requires greater GPS sensitivity during an arrival, it should be no surprise that the controllers expect the same of you as you depart. Therefore, as a last item on your takeoff checklist, set the GPS to approach-arm mode. You can revert to normal sensitivity after departing the 30-mile radius area of the airport.

Once your GPS is set up, there is no further need to push a button or twist a knob in order to get it going. As long as you have the proper flight plan activated, the unit will sense the movement of the aircraft on its own and will begin waypoint sequencing for you. But watch out! Many people get wrapped up in gawking at the GPS unit during a busy, critical phase of flight such as departure. First fly your airplane—period!

A normal departure from a tower-controlled airport may well carry you away from the desired track to your first en route waypoint—especially if your desired track is perpendicular to the runway departure corridor. Don't be distracted by this.

You have been given the initial heading for your climb-out. So, focus your attention on the gauges as you climb to your assigned altitudes on your assigned headings. Don't fly through an assigned altitude because you are watching the GPS screen! Just as cell phones distract drivers, our exciting new "eye candy" in the cockpit can be just as distracting—maybe even more so with some of the more vividly colored displays.

If you are departing from a major airport, such as Charlotte's Douglas International, the control tower may switch you to a takeoff runway you didn't plan for. Regardless of the circumstances, change your GPS departure procedure, if necessary, to match the new assigned runway. If you depart in VMC conditions, proceed as directed, take a vector or two, make sure you are clear of obstructions visually, then request "direct-to" to get to your first en route waypoint. You will probably get what you want if everything is going along smoothly.

But if the weather is anything less than "severe clear," don't be too quick to request direct-to your first en route waypoint. For there is a major trap awaiting the unwary GPS user in going direct-to anything too soon after takeoff. One of the limitations of GPS is that it can paint a direct line sometimes where it will get us into trouble—namely, too close to the ground near mountains, or too close to tall buildings or hard-to-see cell towers.

Here's how the real world of IFR might set a trap for the unwary: Let's say that the world's least brilliant 200-hour instrument pilot is departing from Birmingham International (KBHM), AL, an airport he is not familiar with. On top of that, he hasn't looked at the Take Off Minimums and Departure Procedures page that covers Birmingham. He doesn't notice that a climb to 2,100 feet on runway heading (runway 24) is required before turning left on course to avoid some antennas lining the ridge to the south of the airport. The tallest of them tops out at 2,049 feet MSL (mean sea level). (See Fig. 6.5, next page.)

In his Piper Seneca, he is heading for Mobile Downtown, Alabama, also known as Brookley Field (KBFM). His GPS is set up to take him there direct-to. Shortly after entering clouds at 300 feet AGL, he loses communications. In the panic that ensues, he forgets his initial climb clearance and decides to rely solely on his GPS. He turns south to join the direct course depicted on the screen. Will his luck run out, or will he unknowingly thread his way through those four tallest antenna towers?

The same hazard can occur during missed approaches at smaller airports when the GPS draws a straight line to the holding fix.

Normally what happens as you depart from a smaller airport is that you fly runway heading, or get a slight turn, and are told to contact departure control. When

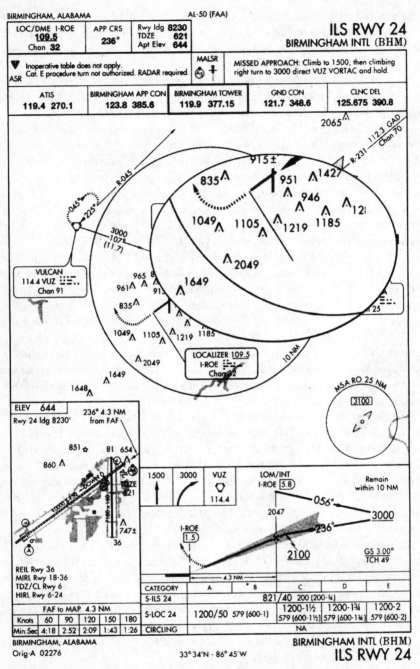

Figure 6.5 Obstructions are magnified on this approach chart to show danger of not following departure instructions on takeoff at Birmingham International, AL.

you contact departure after your climb is established, you will usually continue straight ahead or continue the slight turn. Departure control's instructions guarantee you obstacle clearance.

Moving Map

For most of us, having the large, moving map of a GPS multifunction display (MFD) is the best situational awareness tool ever invented. You already get excellent, precise horizontal guidance through your CDI or HSI when it is fed by GPS output. Adding a large moving map to the GPS feed allows you to cross-check your other navigation at a glance and see exactly where you are in relation to where you should be.

The amount of detail is surprising, even on the smallest display. With the moving map, you can easily scale down to a mile or less in the vicinity of a busy airport—where it counts the most. If you need to look out over a greater area, you can expand the scale up to a comfortable 30 nm or more.

At what point do you bring up the moving-map display? Why not check it out as a part of your GPS start-up routine? That way you can set the range scale and other options and have the map operating the way you want it when you taxi out. Set the range scale at 1 nm for takeoffs, approaches, and other close in work around an airport.

But BEWARE! The moving map is probably the most attention-grabbing, mesmerizing, fixating piece of equipment ever introduced into an airplane cockpit. You will have to make a conscious effort at first to avoid fixating on that fascinating display. Your eyes will naturally be drawn to something moving, especially if it is vividly colored. Include it in your instrument scan as you would any other navigation instrument. A quick look from time to time is all you need to check your progress.

Equipment Note: On the Bendix/King KLN 94, the moving map is displayed by activating the NAV 4 page. The Garmin 530 displays the moving map through the NAV 2page. The Apollo GX60 is accessed through the MAP button.

Distractions

If at any point ATC gives you a clearance that will distract you too much from your instruments during rough weather or a mechanical or electrical problem, you should immediately tell ATC that you are unable to take it. ATC will then have to offer another option that should help you out. If your workload is high, the ATC controllers will do their best to accommodate you. If you give them a little insight as to why their clearance makes your situation difficult (e.g., you're being thrown around by a bad pocket of turbulence), they'll understand. They know, as you know, that flying the airplane always comes first. Be as assertive with the controllers as the situation merits. Accidents have occurred when a controller hasn't taken the hint from the pilot that he or she is in a potentially dire situation.

En Route Operations

When you reach a safe altitude, ATC will probably clear you to "resume own navigation." ATC now expects you to proceed directly to your first en route waypoint. Now you can turn to your GPS, highlight the first en route waypoint on your flight plan page, click direct-to, and then enter.

When established on course, ATC might offer you "cleared direct destination." If ATC doesn't make this offer, you may request it. Whether you are in the middle of a departure procedure or simply heading to your first en route waypoint, go to your flight plan page and scroll down to highlight your destination.

If you're planning a VFR approach to that airport, your destination is the airport itself. If you are expecting an IFR approach, highlight the initial approach fix (IAF), press direct-to, and then enter. It would be premature at this point to assume you'll get vectored to the final approach course.

Messages

When the MSG annunciator lights up, it's the GPS equivalent of your Internet service provider's "you've got mail!"—except that GPS messages are much more important than a good many e-mails. For this is the *only* way you are notified about problems discovered by RAIM (random autonomous integrity monitoring).

If all is well, RAIM operates in the background, and you never know it's there. But if RAIM senses anything wrong or out of limits among the satellites providing data for your flight, it will alert you with a message. This is the GPS equivalent of seeing red flags drop on VORs and ILSs.

If a message pops up with a RAIM problem en route, you must switch over to your VOR backups, discontinue navigating by GPS, and notify ATC of the change in your situation. You can continue to monitor GPS and hope the problem goes away.

But if you have commenced an approach and are inbound from the LAY, you should notify approach control immediately that you have a problem and request a non-GPS approach, or if VFR, cancel IFR with approach control and tell them what your intentions are. If you execute a missed approach and head for your alternate, you must be prepared to use a non-GPS approach at the alternate.

We will discuss RAIM emergencies in detail in Chap. 8, Outages, Emergencies, and Other Surprises.

There are dozens of other messages programmed into your GPS. Some relate to non-RAIM technical problems, and others concern the progress of the flight, such as an "Airspace Alert" warning that you are about to enter a special use airspace. Browse the sections of your manual that list messages, and mark the beginning of these listings so that you can quickly turn to them in the cockpit. You do carry your GPS manual on all flights, don't you? It's required whenever there is a GPS unit installed in a plane.

Or simply stop using the GPS and navigate with your backup VORs if a message appears that you don't understand.

VOR and GPS Differences

One thing you'll find flying en route with GPS is that the VOR airways will not always match up with what your GPS says.

For example, a Victor airway may be defined by the 123° radial from the ABC VOR. Meanwhile, you are flying what you think is the equivalent on your GPS, and you find your GPS track is a mile or two offset from the VOR airway. Which one is correct? Of course, it's the track that ATC observes as being correct.

Since the width of a VOR airway is four nautical miles on either side of the airway centerline, your GPS track may be within parameters. As for which one is the most accurate, our money is on the GPS. The newer, pinpoint GPS technology far surpasses the accuracy of the VOR system, which was designed in the middle of last century.

Being optimistic, let's hope you're cleared direct-to your destination with GPS as your primary navigational resource. Quite possibly this part of your trip will afford you time to calculate your fuel burn and match it with your planning estimates, and also to look over the destination approach plate one more time.

An Airline Technique You Can Use

Some of the airlines that use GPS primarily for en route operations actually file a flight plan based on Victor airways. Once they are well on their way, the pilots request direct-to their destination and switch over to GPS to navigate the rest of the trip.

They continue to use the flight plan/nav log constructed with Victor airways and VORs to keep up with their fuel usage. With GPS feeding NAV 1, they set up their NAV 2 on an upcoming VOR station and then set the OBS to match their course. They use the NAV 2 station passage for fuel readings and time estimates. Equally important, the VORs serve as backups in the event of a GPS outage or malfunction.

En Route Clearance Changes

"I have an amendment to your routing," says the ATC controller. "Advise when ready to copy."

"Ready to copy," you reply. But what you're really thinking is, "Oh heck!" or words to that effect. So, first things first. Write down the clearance accurately and read it back to the controller. Click over to the active flight plan page, if you're not there already. If the clearance is to navigate immediately to a new waypoint, cursor down to the waypoint past your current active waypoint. Punch in the new waypoint, verify that you entered the right data, push DIRECT-TO, and then ENTER.

This will immediately take you to the new waypoint. Since you have bought yourself a little bit of time, stop and take a look at the flight instruments and engine gauges. Look around outside the cockpit. Make sure all is well. Now, spend a few moments entering the rest of the new clearance.

Using the cursor again, add the next waypoints in order behind the active waypoint. Delete the unnecessary waypoints. Needless to say, look up from time to time and make sure the airplane hasn't rolled over on its back.

If it becomes necessary to change the arrival procedure you were originally cleared for, go into the procedures (PROC) function of the GPS and bring up the page listing all of the arrival procedures for the destination airport. Highlight your new arrival procedure and enter. You will be prompted with the question, "Do you want to replace the previous arrival procedure?" Approve this by clicking OK or ENTER. Next, you will be asked for a transition. Choose the transition you want and enter that. Now you are "good to go."

Occasionally, you will be cleared to a point well inside the arrival procedure. But your choice of a STAR does not include a transition for that particular waypoint. In this case, choose any other transition that includes the cleared waypoint and click OK. Then use the direct-to function to choose the appropriate waypoint. Set up a direct route to that point.

Arrival, departure, and approach procedures each come in clusters of waypoints that are preset. By this we mean that the user cannot change waypoints inside these clusters. The FAA required GPS manufacturers to set up these procedures in this manner to keep people from inadvertently adding erroneous waypoints close to airports.

Diversions

One situation that occurs quite often when you're en route is the need to deviate around a thunderstorm or other bad weather up ahead. ATC will grant your request for a deviation around the weather if at all possible and will tell you to "resume own navigation" once you are past the danger. You set the heading bug, turn to the new heading, and go around the weather.

Depending on how far off course you have to deviate, you then have two choices: You can request ATC clearance direct-to your next waypoint beyond the bad weather, or—if the deviation was slight—you can create your own intercept course to get back on the GPS track without resetting the direct-to feature. In most cases, it's probably best to ask ATC for clearance to go direct to the next waypoint. This provides some insurance that you and ATC are on the same wavelength

Nearest Function

How would you handle a scenario where the weather ahead becomes too heavy to deviate around? For this and other scenarios—such as emergencies—you may want to land as soon as possible at an airport beside or behind you. In this case, GPS has a function that no other system provides—the NRST (nearest) feature.

The NRST airport function instantly accesses the closest 20 or so airports with their frequencies, runway lengths, and other important data at your fingertips. (You can also select the nearest VORs, NDBs, intersections, user-defined waypoints, special use airspace, FSS frequencies, and ARTCC frequencies.)

As you scroll through the nearest airports, you will see the closest one first along with the heading and distance to that airport. Also within a finger poke are all the approaches down there. Simply select the airport you want with the NRST button and ENTER.

En Route Holding

Now for your next challenge: "November one two three four, you are cleared to the foothills VOR, hold northeast on the zero six zero degree radial, right turns. Maintain six thousand feet. Expect further clearance at eighteen hundred Zulu."

You have kept ahead of the curve during this flight; so you already have the latest weather for Dekalb-Peachtree airport. A heavy thunderstorm is rumbling across the airport. There are three aircraft in the holding pattern ahead of you. Glancing down at the fuel gauges, you see that you have over two hours of fuel remaining in your tanks. So you can probably hold until the thunderstorm moves on.

Now for the hold. This is where there are many differences between the units now in use. Some fancier installations even have a display that will give you a bird's eye view of the holding fix, along with patterns for missed approach or course reversal holds. But ours is a simpler piece of equipment. Let's see how it can help us with holding in our current situation.

First of all, is the holding fix in your new clearance one of the waypoints already in your flight plan on the GPS? If so, just keep on until you get there. If not, dial up your holding fix with the cursor. Once that is done, push the DIRECT-TO button and ENTER to give you a direct course to the holding fix.

Next, let's set up the GPS so that it acts like a VOR. On your CDI, push the OBS button. Using the appropriate knob, scroll around to the course you are to hold on (in this example, 060°). Pushing the OBS button removes the autosequencing function so that when you pass your waypoint your GPS stays referenced to that waypoint instead of autosequencing to the next waypoint. In the cockpit, you'll see the GPS go from a "to" to a "from" indication as you pass over the waypoint.

You will now fly the CDI or HSI as a VOR hold. The only difference is that GPS is providing the guidance. You can also monitor your progress by going to the GPS moving-map function and adjusting the range scale to focus on the holding pattern.

This is important: When you leave the hold, don't forget to push the OBS button again to resume autosequencing of the waypoints for the approach.

As you head for the airport, your main concern is staying ahead of the airplane. You must get the weather from ATIS, brief the approach plate, run the appropriate checklists, and set up your GPS for the appropriate arrival procedure. You'll be wishing for an autopilot if you don't already have one.

VNAV Can Be Distracting

VNAV allows the pilot to set up the GPS to alert him or her when it is time to start a descent or climb to reach a desired altitude. The pilot enters starting altitude, feet per minute desired, level-off altitude, and distance in miles from the level-off point. The box does the rest. An alarm signal or the message annunciator flashes at the calculated point to begin descent or climb based on the aircraft's ground speed.

While this is a neat and fairly simple function, it is more often an unnecessary crutch or, sometimes, a huge distraction. During a VNAV descent, new GPS pilots frequently find themselves measuring their progress against the VNAV readout. This can easily distract the pilot from instrument or outside scanning. It also may lead to "throttle jockeying" and/or chasing the readout with pitch oscillations, a form of overcontrolling that is not very professional or comfortable for anybody on board.

Slower aircraft rarely need VNAV. If you need to lose 6,000 feet to reach your target altitude for an approach, that's 12 minutes at a descent rate of 500 feet per minute. At 60 knots ground speed, you will cover 12 nm in 12 minutes; so begin descent 12 nm out. At 90 knots ground speed, you will cover 18 nm in 12 minutes; begin descent 18 nm miles out. At 120 knots, start 24 nautical miles from the airport, and so forth. You can do these calculations in your head without getting distracted from your instrument or outside scan.

Turboprops and jets, on the other hand, like to stay high until the last possible minute. For fuel economy, a King Air of Lear pilot may chop the power and descend at 2,500 or 3,000 feet per minute. Add to that the complexities of calculating crossing altitude restrictions imposed by ATC and you will find these pilots love to have their in-the-head calculations cross-checked by the GPS VNAV function.

When you get used to using GPS—or if your workload is small—help yourself to the VNAV feature. This is a great function to practice when you make altitude changes on a VFR cross-country flight.

Black Box Practice-4

1. Create your own personal checklist for the start-up checks your particular system runs through when you turn it on. Add checklist items for fuel on board, nearest preferences, enabling turn anticipation, and moving map options with the specific steps needed to bring up the pages for these items.
2. Enter the flight plans your instructor assigns, including DPs, STARS, and approaches when available.
3. Store these flight plans and practice retrieving them.

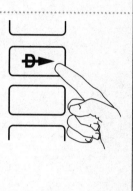

7

GPS Approaches

Let's take a tour through any NOS booklet of approach charts and see what GPS approaches we can find. The results of this tour may surprise you. For you will quickly see exactly where we are in the transition from a ground-based to a space-based air navigation system. And how much farther we have yet to go.

The first thing you will notice is how many conventional VOR, NDB, ILS, and other ground-based non–GPS approaches are still available. For example, as this is written, Charlotte Douglas International, NC, the departure airport discussed in Chapter 6, lists eight ground-based, non-GPS approaches altogether, not counting ILS Category II and III. However, Charlotte has only six GPS approaches.

This will change with time, of course, as the FAA continues its transition to a space-based system. Meanwhile, we must remain proficient in ground-based approaches *and* master the new space-based GPS approaches. We must be able to pass check rides based on *both* systems.

GPS Overlay Approaches

Next, notice how many GPS overlay approaches there are. Overlays, as the name implies, are GPS approaches laid down over existing ground-based navaids. They are flown the same way except that course guidance comes from GPS in the space-based procedure and not from the ground-based navaids in the underlying procedure. Their dual nature is indicated in their designations, for example: VOR or GPS RWY 22, NDB or GPS RWY 2, and so forth. See Fig. 7.1 (next page) for a typical overlay approach laid down over a VOR approach. This is the VOR/DME or GPS RWY 20L approach for Dekalb-Peachtree, GA, the destination for our hypothetical cross-country flight.

In the early 1990s the FAA chose to introduce the flying world to GPS approaches with overlay procedures. This not only cut the production time of new approach charts but also allowed pilots to go from what they knew to something new. Overlays gave pilots familiar backups as they executed the new GPS approaches.

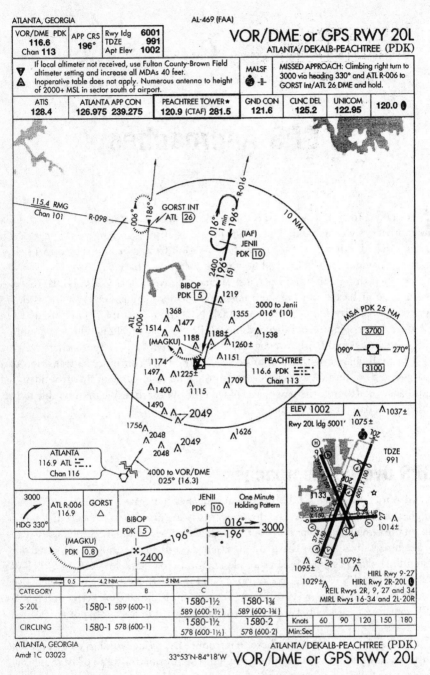

Figure 7.1 GPS RW 20L at Dekalb-Peachtree, GA, is a GPS overlay approach based on the VOR/DME 20L.

But some of them can be pretty interesting at first glance! See Fig. 7.2 (next page), the NDB or GPS RWY 17 at Corning, IA, and Fig. 7.3, the VOR or RWY GPS 35 arc approach at Paris, TX (page 79).

Overlay approaches were never intended to stay around forever. The FAA has created its last overlay approaches, and existing overlays are being converted to GPS RNAV procedures. So overlays will still be with us for a while as they are changed over to the RNAV format.

GPS Stand-Alone Approaches

The next type of GPS approach that you should note in the NOS booklets is the "stand-alone" procedure. You may have to hunt for the stand-alones because they are not as numerous as overlays or RNAV approaches. Stand-alones, as their name implies, are not dependent on ground-based navaids.

Stand-alones are designated simply as GPS RWY 6, GPS RWY 25, and so on. The non-RNAV stand-alone is frequently associated with small airfields or those that present special challenges. See Fig. 7.4 for a good example of the stand-alone approach, the GPS 19 at Ahoskie, NC (page 80).

While you are looking at Fig. 7.4, check out Fig. 7.5, the Basic T design as shown in AIM (page 81). Notice the similarity of the patterns to the Basic T. All GPS approaches—new ones as well as older ones being upgraded—will eventually conform to the Basic T. This is very good news. The Basic T is simpler to understand and easier to use. It can be applied to the smallest airports as well as to the largest and busiest. This is a degree of approach standardization we have never seen before.

GPS RNAV Approaches with WAAS

Now let's examine the GPS RNAV approach with WAAS minima—the ultimate direction in which all GPS approaches are evolving. See Fig. 7.6, RNAV (GPS) RWY 17R approach for Oklahoma City–Will Rogers World (OKC) on page 82. This chart is a good example of the new features being introduced with the shift to GPS RNAV and the lower approach minima available with WAAS.

The next approach chart (Fig. 7.7 on page 83) is the RNAV (GPS) RWY 23 for Frederick, MD, headquarters of AOPA. This approach is typical of what is becoming available for smaller airports, replacing GPS overlays and stand-alones.

At the top of the chart, the "pilot briefing" boxes have been expanded to provide better quick references while setting up and flying an approach. Most of the items are familiar but are now presented in a format that makes them easier and quicker to refer to.

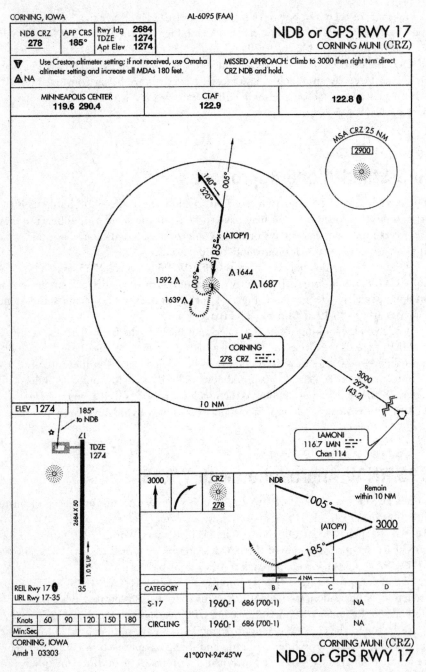

Figure 7.2 GPS overlay approach at Corning, IA, with all approach and missed approach fixes located at the NDB on the field.

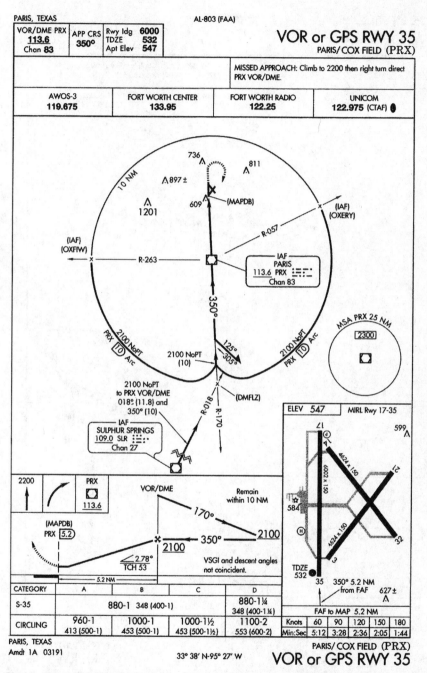

Figure 7.3 This DME arc approach to Paris, TX, is much simpler with GPS. Instead of resetting the OBS every 10°, simply keep the OBS needle centered with small heading changes as the GPS guides you to proceed around the arc.

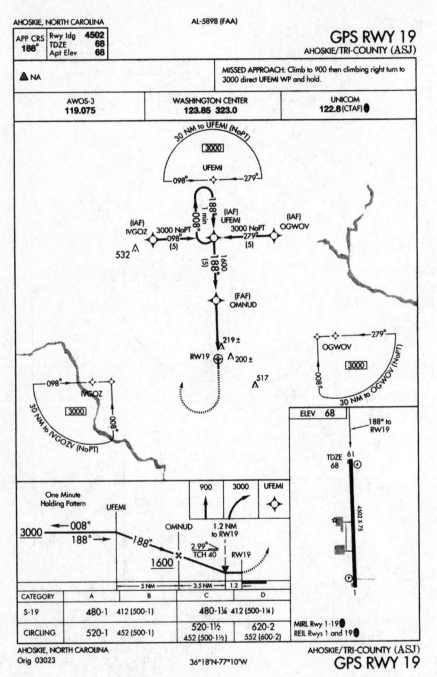

Figure 7.4 GPS stand-alone approach to Ahoskie, NC. Compare with Basic T design shown in Fig. 7.5.

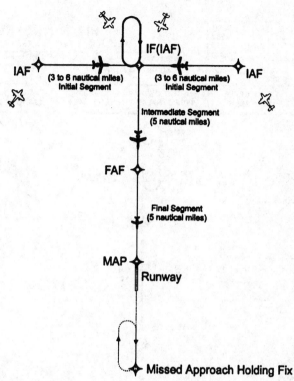

Figure 7.5   The Basic T design for GPS approaches. With minor adaptations, this design is now used for all new GPS nonoverlay approaches.

New Approach Chart Items

There are some new items, however, that are important to understand for GPS RNAV approaches. Let's take them from the top down, starting with the top left corner of the Oklahoma City chart (page 82):

1. *WAAS Ch 00500.* This is a "WAAS Channel Number," an optional equipment capability that allows the use of a five-digit number to select a specific final approach segment. Below this you will see W-17A, a five-digit identifier for verifying selection of the correct final approach segment. Nothing has been published, as this was written, about how these numbers are going to be used.
2. *BARO-VNAV NA below 17°C (2°F).* Barometric altimeter minima are not applicable for RNAV approaches when the temperature drops below 17°C (2°F). In other words, it is illegal to make RNAV approaches using only barometric altimeter inputs when it gets this cold.

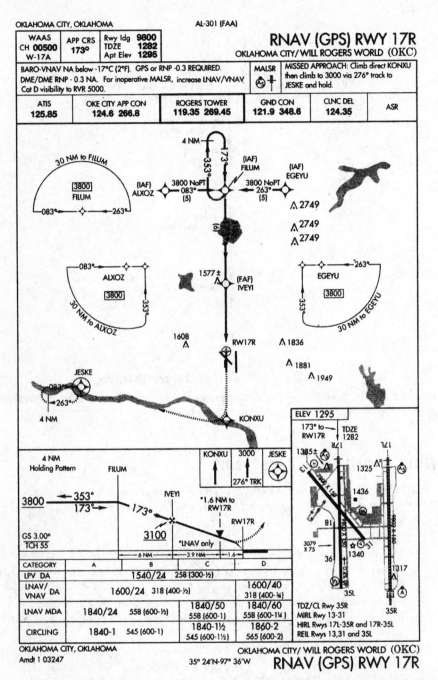

Figure 7.6 The Will Rogers World airport at Oklahoma City, OK, received one of the first seven WAAS approaches issued by the FAA. The WAAS-based LPV minimum for RWY 17R is 1540/24 for light planes, compared to the ILS minimum of 1482/24 for the same runway.

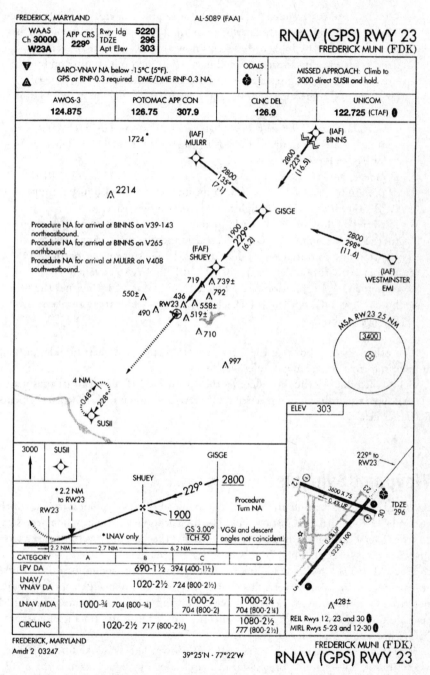

FREDERICK, MARYLAND   AL-5089 (FAA)

WAAS Ch 30000 **W23A**	APP CRS **229°**	Rwy ldg **5220** TDZE **296** Apt Elev **303**

RNAV (GPS) RWY 23
FREDERICK MUNI (FDK)

▼   BARO-VNAV NA below -15°C (5°F).
▲   GPS or RNP-0.3 required. DME/DME RNP-0.3 NA.

ODALS

MISSED APPROACH: Climb to 3000 direct SUSII and hold.

AWOS-3 **124.875**	POTOMAC APP CON **126.75** **307.9**	CLNC DEL **126.9**	UNICOM **122.725** (CTAF)

FREDERICK, MARYLAND
Amdt 2   03247

39°25'N · 77°22'W

FREDERICK MUNI (FDK)
RNAV (GPS) RWY 23

Figure 7.7   Frederick Municipal airport in Maryland also received one of the first seven WAAS approaches. The LPV minimum for RWY 23 is 690-1 1/2, compared to the ILS minima of 598-1 for the same runway.

3. *GPS or RNP 0.3 required.* RNP stands for required navigation performance. This concept was recently developed by the FAA to ensure that the performance of the many high-tech systems now available have the precision needed for the new approaches and navigation methods coming on line. In this example, your equipment must have an RNP rating of 0.3 to legally (and safely) fly this approach. Properly certified panel-mounted GPS units have an RNP of 0.3.

4. *DME/DME RNP 0.3 NA.* For whatever the reasons, DME/DME RNAV approaches are not available for this runway even if the DME/DME equipment has an RNP rating of 0.3.

5. *For inoperative MALSR, increase LNAV/VNAV Cat D visibility to RVR 5000.* High-speed aircraft in Cat D must increase the visibility minimum to RVR 5000 if the MALSR lighting system is inoperative.

6. If a briefing box in the upper left panel contains this symbol: **W**, then there may be outages of WAAS vertical guidance occurring daily at this location due to initial system limitations. Refer to page A1 of any NOS approach chart booklet for further details. As WAAS coverage is expanded, **W** limitations are expected to be removed. (Neither the Oklahoma City, OK, nor the Frederick, MD, approach chart examples shows the **W** symbol.)

The other briefing boxes at the top of an NOS approach chart booklet present familiar information in a quick reference form.

Below the pilot briefing boxes comes the "planview" of the terminal arrival area (TAA). Note that minimum safe altitudes (MSAs) are shown in pie-shaped segments for better clarity.

Waypoint Symbols

Fixes and other checkpoints are marked mainly by "fly-by" waypoint symbols, which also indicate that turn anticipation should be used at these locations, when necessary, to avoid overshooting the desired track. Note that the 17R approach has three IAFs (initial approach fixes). This is the Basic T arrangement.

The IAF at FILUM has a holding pattern in lieu of a procedure turn if a course reversal is needed. We have seen this many times before in non-GPS approaches, but note that the outbound leg here is "4 NM," not the usual one minute. The 4 NM turning point gives the GPS additional time and distance to "turn itself around" before proceeding inbound to the IAF.

The missed approach holding fix at JESKE (lower left) is marked by a different waypoint symbol—a four-pointed star in a circle. See Fig. 7.8. This indicates a "fly-over" waypoint. You must fly over this type of waypoint before beginning a turn to a new track. Turn anticipation is not available for fly-over waypoints.

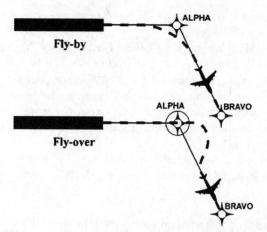

Figure 7.8 At fly-by waypoints (top) turns should start before waypoint using turn anticipation. Fly-over waypoints are used when the aircraft must fly over the point before starting a turn.

Farther down, the "Profile" and "Airport Diagram" sections contain no major changes. If you are uncertain about some of the symbols or other information, review sections F through L in the front of the NOS booklet.

But pay attention to the dark "V" just before the missed approach waypoint. This indicates a visual descent point (VDP). The pilot should not descend below the MDA prior to reaching the VDP even if past the FAF with the field in sight.

RNAV Approach Minima

The "minima" section at the bottom of each RNAV approach chart plunges us into an unfamiliar alphabet soup. The aircraft categories—A, B, C, and so on—are the same as always. But new approach categories—LPV, LNAV/VNAV, and LNAV—have been introduced to accommodate WAAS approaches. Let's take them from the top:

1. *LPV DA.* This is the line for WAAS approaches with lateral and vertical guidance. Despite its appearance, LPV is not strictly speaking an acronym, but is more a concept identifier. Think of LPV as a WAAS-supported approach with **L**ateral **P**recision equivalent to present-day localizers and **V**ertical guidance provided by a WAAS-generated electronic glideslope. LPV replaces GLS on this line. GLS, or global landing system, will accommodate aircraft equipped with precision approach certified WAAS receivers operating at their fullest capability. But no timetable has yet been established

for its implementation. Current GLS NA lines are being removed and replaced with LPV lines.

2. *LNAV/VNAV.* This stands for Lateral Navigation/Vertical Navigation. LNAV/VNAV approaches have minimums based on GPS lateral precision plus an electronic glideslope. LNAV/VNAV approaches permit lower minimums than nonprecision approaches. Altitude guidance for the glideslope can be provided by either electronic or barometric inputs. Since these altitude inputs generate an electronic glideslope, the LNAV/VNAV minimums are listed with a DA.

3. *LNAV.* This stands for Lateral Navigation, the designation for nonprecision approaches to an MDA with no electronic altitude capability, only, lateral guidance.

To summarize the approaches now coming on line with WAAS:

WAAS-enhanced GPS can handle all three of RNAV approaches listed. LPV approaches can only be made with WAAS.

LNAV/VNAV approaches can be made with WAAS and also without WAAS by using GPS for lateral guidance and baro-VNAV coupled to an electronic glideslope for vertical guidance.

LNAV approaches can be flown both WAAS and GPS, but have only lateral guidance. They provide no electronic glideslope for vertical guidance.

GPS Approach Basics

The essentials of flying a GPS instrument approach are straightforward and familiar—you intercept and follow the track indicated on your CDI or HSI. You descend as shown on the approach chart as you reach the various waypoints until you reach MDA or DA. Then you land or execute a missed approach.

The big difference with GPS is that you are working with a computer for track and waypoint information, not fixed signals from the ground. The choices offered by your GPS computer are almost unlimited. So you must enter your key strokes carefully, or your GPS might start guiding you to someplace you don't want to go.

Setting up your GPS for an instrument approach begins before takeoff from the departure airport. Ask yourself which approach will most likely be in use when you get to your destination, based on your analysis of the weather. Then enter this approach as the final step in loading your flight plan into your GPS unit before takeoff.

In Chap. 4 we suggested that you develop your own do-it-yourself checklists to help you master the keystrokes for the various functions used on a GPS flight. Do-it-yourself checklists can be tailor-made to provide exactly what you need to remember for each phase of flight, along with the keystrokes you need to enter for your particular make and model GPS.

A do-it-yourself checklist for selecting, entering, and activating approaches can simplify your work when you set up your flight before takeoff. This checklist can also help reduce the confusion (panic?) in flight that comes with guesswork and trial-and-error stabs to correct wrong moves during a busy approach.

Here's what a do-it-yourself checklist might look like for selecting and entering approaches with the Garmin 530:

SELECTING AND ENTERING APPROACHES

1. PROC key: Press to display Procedures page.
2. Rotate large right knob to highlight "Select Approach."
3. Press ENT.
4. Rotate large right knob to highlight desired approach.
5. Press ENT.
6. Rotate large right knob to highlight "Load?"
7. Press ENT.

TO ACTIVATE LOADED APPROACH:

1. PROC key: Press to display Procedures page.
2. Rotate large right knob to highlight "Activate?"
3. Press ENT.

Setting Up the Approach

When you are 40 to 50 miles out from your destination, tune the destination ATIS. It will give you the current weather, altimeter setting, instrument approach, landing runway, and field conditions.

Maybe you'll get lucky, and the approach in use will be exactly what you expected and entered before takeoff.

But don't count on it. If the approach has changed from what you planned, reshuffle your approach charts, refresh your memory with the pilot briefing boxes and altitude minima on the new approach, and enter the new approach on your GPS. Make sure your non-GPS VOR, NDB, and/or ILS backups are also correctly set as backups for the GPS approach.

Think and plan for *all* possible approach scenarios. Mentally work your way through each scenario and decide ahead of time how you will handle it, including the specific keystrokes for making the GPS do what you want it to. If you go through this mental drill on *every* flight you will find that approaches will soon become routine.

There are four basic scenarios for flying an instrument approach. They apply to GPS as well as non-GPS approaches. So you have probably been rehearsing these scenarios since the day you first encountered approaches in your training for the instrument rating. Here are the four approach scenarios you must be prepared for as you near the end of your IFR flight:

- The approach you planned
- A different approach indicated by ATIS or assigned by ATC or approach control during the flight
- Missed approach and return for another pass
- Missed approach and proceed to alternate

As you mentally rehearse these scenarios, always ask yourself

1. What is the best way to set up my GPS and the non-GPS backups for each of these scenarios?
2. Do I have to do anything with the GPS to cope with radar vectors and holding patterns?
3. How do I handle the GPS if approach control changes the approach on me before I get to my IAF? After passing an LAY heading for the FAF?
4. How do I execute a GPS missed approach and set up the GPS to return for another landing?
5. How do I execute a missed approach and set up the GPS to proceed to my alternate?

The Autopilot Advantage

For obvious reasons the autopilot can be a godsend when the time comes to set up GPS approaches and their VOR backups—or make quick, last-minute changes in your plans, such as executing a missed approach and proceeding to an alternate. There is nothing like it to hold the airplane steady as you punch in the numbers. While you must still "aviate, navigate, and communicate" as you deal with your GPS, the autopilot frees up additional time between instrument scans to work things out.

The autopilot is a welcome helper in your cockpit at times of greatest button-pushing activity. Don't be shy about using it to the maximum! On an instrument rating check ride your designated examiner will be impressed if you use the autopilot fully and competently.

Keystroke Briefings

Instructors should brief the events and keystrokes (and autopilot use if installed) on *every training flight that involves an approach,* including the approaches demonstrated by the instructor in the early stages of the syllabus. Match the keystrokes for each event with the specific make and model GPS the student will be using. Then have the student practice the keystrokes you brief on a docking station, simulator, or in the cockpit on the ground until he or she can run through them without fixating too much on the GPS display.

To make your job easier, write out all your keystroke briefings ahead of time. Keep these briefings. They will quickly become a valuable resource. Have your students also write out the main points of the briefings as do-it-yourself checklists. Have them save these checklists for further reference during their training.

Here's the way a keystroke briefing might be written out for missed approaches with the Bendix/King KLN 94:

If unable to land at the MAP commence a normal climb straight ahead and

1. FLY the exact published missed approach procedure from the approach chart. Do not turn until reaching the missed approach turning altitude.
2. CONFIRM automatic selection of the first waypoint of the missed approach procedure (usually the missed approach holding point).
3. Press [**Đ►**] then ENTER.
4. As you approach the holding waypoint, press OBS to suspend waypoint sequencing and setup up your CDI or HSI to enter holding and fly the pattern.

Continue on with this briefing as needed to cover the next scenarios as needed: (1) missed approach and return for another pass and (2) missed approach and proceed to alternate.

Final note to the instructor on approach briefings: As your student becomes more proficient with GPS approaches, have him or her *brief you* on the events and keystrokes for the flight. There is no better way to make sure students are mastering the material than to have them play it back to you. It's a great confidence builder for students. It also prepares them for questions they might get on the oral part of a check ride. There is simply not enough time to cover every detail of every task in flight. So students can count on getting quizzed about any—or all—the tasks required on the check ride.

Scaling

Whenever you load an approach into a flight plan, the GPS will start setting up for that approach at 30 nm out. At that point the CDI sensitivity changes from the en route scale of 5.0 nm on either side of the track centerline to the terminal scale of 1.0 nm.

This means that full deflection of the CDI needle changes from 5.0 rim to 1.0 rim. See Fig. 7.9 (next page). The transition is gradual and easy, as shown in Fig. 7.9, and not abrupt. Just keep the needle centered.

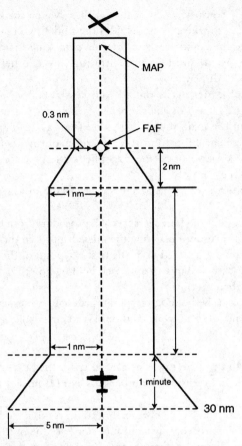

Figure 7.9 The CDI needle scales down from 5.0 nm to 1.0 nm within 30 nm of the destination airport. The CDI then scales down to 0.3 nm during the approach. *(Courtesy of Garmin Ltd.)*

The CDI changes sensitivity one more time as you get closer in. At 2.0 nm from the FAF the sensitivity scales down further to 0.3 nm on either side of the centerline. Sensitivity stays at 0.3 nm until the MAP is reached, where it scales back up to 1.0 if you haven't landed. Annunciators provide alerts as you reach these transition points.

Now look at Fig. 7.10. This illustrates a feature that makes flying GPS approaches much steadier than using ground-based navaids.

You already know how much faster the CDI needle moves from side to side as you get closer and closer to an NDB, ILS, or VOR. That's because of the narrowing of the signal. This doesn't happen when you fly an approach with GPS because the system is guiding you within a fixed distance from the centerline and not down a converging beam.

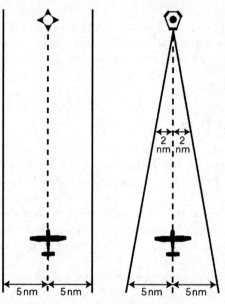

Fig. 7.10 Unlike VOR, the range of GPS needle deflection does not get narrower as you get closer in, but remains constant. *(Courtesy of Garmin Ltd.)*

When you begin working with WAAS approaches, you will see that flying an electronic glideslope is also much steadier for the same reason. The WAAS electronic glideslope is flying you along an altitude slope, not down a fixed beam.

In the Cockpit on a GPS Approach

Knowing what you now know about approaches, let's continue on with the hypothetical flight we began back in Charlotte at Douglas Field. At the point we left off, you had arrived at the fringes of Atlanta's Class B Airspace and were having to hold for thunderstorms in the Atlanta metro area. Those storms have since passed, and Atlanta approach is now able to pick up the pace for the numerous arrivals waiting to land, including you.

One thing has changed, however. The wind has shifted and is blowing a little harder. Hartsfield, the main airport at Atlanta, was landing to the west and still is after the wind shift. But a major airport such as Atlanta's Hartsfield often dictates the arrival procedures at other airports close by. Calling it 310° at 12 knots, Dekalb-Peachtree's wind now strongly favors runway 27 for an instrument approach at Dekalb-Peachtree. You will have to ditch your plans to land on runway 20L. Expect the unexpected, because you'll get it if you don't! (See chart on page 50.)

"November one two three four, fly heading one four zero, vectors for the GPS runway two seven at Peachtree," says the controller in a hurried clip, reflecting the hectic demands of his duties at the moment. You crisply read back the new instructions. OK, now what do you do next, and how do you reconfigure the box?

Approach Vectors

This next sentence may be the most important one in this book: Keep your scan going while being vectored, and fly the airplane. You are about to change approaches, fly with an unfamiliar piece of equipment to an unfamiliar airport, and respond to controller instructions at auctioneer speeds—all while controlling a hurtling hunk of metal known as a rental aircraft down to about 500 feet above the ground.

Do you have an autopilot? Do you trust it with your life? Yes? Use it. Scan the GPS display as you scan everything else, but don't fixate. Moving colors catch our eyes! Fly the plane!

Next, you have two priority items to consider. First, you have to fly the vector heading exactly, and second, you must brief the new approach plate. Quickly. Go to the airport's approaches page on your GPS. You can push the PROC on the KLN 94 or use the knobs on the Garmin products. Choosing another approach will prompt any of the units to ask if you want to replace the existing approach (in this case, the 20L approach). You do.

Systematically read the briefing strip at the top from left to right, top to bottom. Double-check to make sure the GPS/NAV selector switch is set to feed from the GPS. Make sure your VOR backups are set the way you want them, should you receive a RAIM warning during the approach.

After what seems like an eternity, you hear from the controller again. "November one two three four, descend and maintain three thousand, turn right heading one eight zero, this will put you on a base leg." "R-r-roger, one-eighty on the heading, down to three thousand, Cessna two three four."

On a vectored approach, you must make sure the unit and you are thinking alike. You've loaded the new approach, but which waypoint do you make the present GPS navigation waypoint? Well, that's a good question. You're coming in at an angle to the final approach course. If this were a VOR approach, you would have set in a radial to intercept the final approach course. With GPS, if you push direct-to and choose any old upcoming waypoint, you might have to turn *away* from your vectored heading. Think about it.

Even if you were lucky enough to have a waypoint in front of you, your situational awareness will be compromised when the controller gives you an intercept heading and the GPS suggests differently.

So the trick is to select a waypoint in front of you where you will have to begin your final stepdown. This will be the final approach fix. In the case of this approach at PDK, cursor over and select LIPPY. Before you start staring at the changing lines you've created, push the OBS button. Now, depending on how your airplane is set up, either turn the OBS dial on the face of the VOR to align your instrument's final approach course with the final course on the plate or change the OBS by using the knobs on the GPS.

With the needle on your HSI pointing 97° to your right now, you should see the CDI showing a full scale deflection on the top side of the indicator. You're right of the final approach course. Great. Check the map view on the GPS display. It shows you heading south approaching the final approach course with LIPPY and PDK southwest of you.

Are you hand-flying the plane or using the autopilot? It doesn't matter at this point if all is set up right. The autopilot allows you to take a little more time looking around the cockpit. Autopilots can lead you astray, though. A vacuum pump failure can trick your autopilot into taking you for a wild ride. Keep your scan going, including the vacuum pressure gauge, and read the indications to see if they make sense.

Just then, your controller calls your number. "Cessna one two three four, you are four miles from LIPPY; turn right heading two six zero, maintain two thousand five hundred until established; you are cleared for the GPS runway two seven approach at Peachtree."

Before he even finishes, you're in the turn. As you roll level on the 260 heading, you see the needle sliding toward the middle. Now is the time to push the OBS button to allow autosequencing of your waypoints. This changes the unit from the TO/FROM mode to the TO/TO mode. Your instrument's course is already aligned with the final approach course of the plate.

On Final

Check the map on the GPS display for a quick spattering of confidence. The airport is ahead of you now as you turn the airplane to align with the course. You notice the distance readout on the unit coming up on two miles from the final approach fix. At that point, look for an indication on the annunciator panel or face of the GPS that tells you it is in APPROACH ACTIVE mode. The CDI smoothly tightens from 1 to 0.3 miles on either side of the needle.

The beauty of the GPS final course versus the conventional VOR final course is that it is a consistent width from final approach fix to missed approach point. A VOR course tightens into a point the closer you get to the VOR or localizer, assuming the VOR is in front of you.

As you approach the final approach fix, the waypoint switches to the missed approach point, (MAHTY). Any waypoint in parentheses, such as (MAHTY), is a GPS waypoint that added to an approach plate is also used for a conventional ground-based approach. In other words, a pilot using the same chart to fly the VOR/DME approach to 27 at PDK won't use (MAHTY). It is a GPS waypoint only.

Ahhh, there is a little ground contact here and there. It's tempting to go visual at this point. But you must stay on the gauges until you have the visual cues in sight clearly enough to complete the landing. Our hypothetical IFR cross-country will then be over.

Missed Approach

But suppose you can't land for whatever the reason—visibility below minimums, a RAIM problem, a disabled aircraft on the runway ahead of you. First, cross the missed approach waypoint (MAWP) and execute the exact published missed approach. Do not turn until reaching the missed approach turning altitude.

Report the missed approach to approach control and state your intentions either to (1) proceed to alternate or (2) return for another approach. You will receive a new clearance. Carry it out. Don't get caught off guard if the new clearance is non-GPS!

If proceeding to alternate, go [D▸] the first waypoint or fix in the clearance. Set up the rest of the GPS flight including the approach to expect at the alternate. Set up the non-GPS backups.

If you're returning for another approach at the original destination, don't be surprised if you are directed to hold at the holding waypoint. Go into OBS to enter holding and stay in OBS after leaving holding pattern. Stay in OBS while being vectored until intercepting the final approach course with the final approach waypoint (FAWP) ahead of you. Then punch out of OBS and complete the approach.

This has been a general description of what's involved in conducting missed approaches. As you can see, missed approaches and their options may combine many different functions—vectoring, holding, switching in and out of OBS, proceeding [D▸] , new clearances, and the like. And there are many major differences from one GPS system to another in the way the different functions are handled.

So, practice missed approaches on the GPS equipment you use most frequently. Rehearse the keystrokes for the approaches and missed approach options your instructor assigns. And prepare do-it-yourself checklists to help you nail the right moves.

Black Box Practice-5

1. Enter your startup preferences. Then enter the flight plan and approaches the instructor has assigned for this lesson.

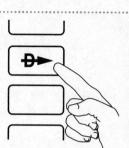

2. Set up and practice the other three other scenarios you might encounter while flying the same approaches as in step 1:
   (a) Change of approach or landing runway
   (b) Missed approach and return for another pass
   (c) Missed approach and proceed to alternate
3. Then enter and practice any overlay, stand-alone, and DME arc approaches in your area similar to those illustrated in the book in this chapter.
4. If your equipment is WAAS certified, practice WAAS approaches similar to the two illustrated in this chapter in Figs. 7.6 and 7.7.

Note: This Black Box Practice covers a lot of ground. Be prepared to spend some time on it.

8

Outages, Emergencies, and Other Surprises

As this book was being written, massive eruptions from the surface of the sun spewed unusually strong bursts of high-intensity radiation at the Earth. This radiation peaked on Wednesday, October 29, 2003. It was strong enough to threaten power grids, communications links, and GPS satellites.

But there were no serious problems with our power grids. They received enough warning, and they were prepared. And, apart from a few scattered instances of radio and cell phone interference, it was business as usual for our communications systems.

GPS continued to function normally—and with good reason. GPS satellites are well hardened against intense radiation such as that from solar flares and nuclear explosions. In addition, the streams of time signals GPA beams toward Earth are analyzed and corrected continuously. Backup satellites orbit with the 24 active satellites, ready to be switched on quickly if needed.

Despite all these features and redundancies, there is always the possibility of a GPS outage of some sort, however remote or far-fetched it might be. And even minor or short-lived problems can affect your ability to complete a flight. In these situations, RAIM becomes very important.

RAIM Warnings

Receiver autonomous integrity monitoring, or RAIM, is available only with panel-mounted GPS systems, not with handheld units.

Let's run through a quick review of the RAIM function, then move on to the steps you should take to handle problems brought to your attention by RAIM.

As an independent function within the GPS receiver, RAIM continuously samples and verifies the integrity of incoming satellite signals. It warns if data from any satellites are missing or too far out of line to provide accurate position information. Unlike the warning flags that pop up when VOR and ILS are unusable, RAIM operates independently of ground-based facilities.

Also, RAIM monitors one satellite more than the number required for accurate GPS navigation. For example, if four satellites are required for an accurate GPS fix, RAIM monitors five. If the signals from one out of the five satellites are substantially inconsistent with signals from the other four—or the required number of satellites cannot be found—RAIM will indicate a problem.

The message annunciator (M or MSG) will light up to alert you if an important message has come through. This message should be read immediately. Some typical RAIM messages that run on three widely used systems:

> *RAIM Position Error.* One of the satellites is too far out of position to provide reliable data (Bendix/King KLN 94).
>
> *RAIM not available from FAF to MAP waypoints.* Sufficient satellite coverage does not exist for the protection limits of this approach segment (Garmin GNS 530).
>
> *En Route GPS RAIM Not Available.* RAIM is not available for en route monitoring (UPS Apollo GX60).

When the RAIM integrity monitor reports a problem, it is telling you that the monitor cannot detect enough fully functioning satellites to guarantee the accuracy of your GPS position within 2.0 nm during en route and oceanic operations, 1.0 nm during terminal operations, and 0.3 during approaches.

So what do you do when you get a message that indicates RAIM is not available or has a problem with a satellite?

You must take immediate action when you get a RAIM warning. As *AIM* emphatically states: "Without RAIM capability, the pilot has no assurance of the accuracy of the GPS position."

Coping with a GPS Outage

Your first move is to check that your backup navaids are set up correctly. You have been monitoring your backups all along, haven't you? Now you know why. If RAIM is telling you it cannot guarantee accurate positions with GPS, you have no choice but to switch over to the ground-based navaids you worked out when you planned the flight.

But don't abandon your GPS too quickly if you are still in the en route phase of your flight. Most GPS outages are transient and short-lived; they are glitches, not catastrophes. CFI Robert Jex put it this way in the November 2003 *AOPA Pilot:*

The few real-world RAIM alerts that I have experienced in more than 1,500 hours of daily GNS 430 use have all been short-lived, with the GPS resuming normal displays and functions after only a few seconds. I would therefore wait a moment to verify the condition before taking alternative measures. The alert may soon clear itself out.

If the RAIM alert does not disappear within a minute or so while en route, ignore the GPS course guidance being displayed. It's not accurate anymore, and there is no way you can fly the great circle route you filed without GPS (unless you can switch to LORAN or INS).

If the RAIM alert persists, you should switch from GPS to NAV and fly direct to the next VOR or other fix ahead of you on your non-GPS backup plan. Tell ATC what you have done, and amend your flight plan accordingly.

Should you switch back to GPS if the RAIM alert disappears? Yes, if you are still in the en route phase of the flight, and the outage is transient and lasts a minute or less. No, if the outage lasts longer, and you have switched over to non-GPS navigation and have amended your flight plan with ATC.

It's a slightly different story if you are in the terminal or approach phase of the flight.

When you are within 30 nm of the destination airport, your GPS receiver will switch from the en route mode to the terminal mode, and the approach you have selected will be armed (some older receivers require that you do this manually).

The receiver runs RAIM predictions by 2 nm prior to the FAWP to ensure that RAIM is available at the FAWP as a condition to entering the approach mode.

"If a RAIM failure/status annunciation occurs prior to the final approach waypoint (FAWP), the approach should not be completed since GPS may no longer provide the required accuracy," AIM states.

At 2 nm from the final approach waypoint (FAWP), RAIM will change to approach sensitivity.

"If the RAIM flag/status annunciation appears after the FAWP, the missed approach should be executed immediately," says AIM.

If a RAIM warning does not appear because of equipment failure after the FAWP, the receiver is allowed to continue without an annunciation for up to 5 minutes to allow completion of the approach.

GPS is basically so reliable that if you start picking up RAIM alerts any time after getting established on the final approach course you know something serious is going on. Abandon that GPS approach.

Check the backup ILS, VOR, or NDB approach you have been monitoring. If you are on course and on altitude to complete the non-GPS approach without excess maneuvering, proceed with it and tell approach control of the change.

If you cannot transition easily and quickly to the backup non-GPS approach, execute a missed approach and decide what you want to do—try another approach at your destination or proceed to your alternate? Remember that you had to file an alternate with non-GPS approaches in case a RAIM alert was due to a GPS problem throughout a wide area.

RAIM Predictions

It is possible to avoid some of these complications as you near the end of your flight by doing a RAIM prediction with your own equipment ahead of time. All you need to do is select the appropriate screen, enter the waypoint you want for the prediction, and a date and time window. You get back a reading on the availability of satellites for those settings.

For example, you can get a RAIM prediction for your destination and ETA before you file a flight plan for GPS training with your instructor. Why bother to depart on a GPS training flight if the satellites won't be there?

Although the function is the same for all GPS receivers, each manufacturer does it differently. Look up "RAIM prediction" in the manual for the equipment you are using, and write up the steps in a do-it-yourself checklist to help you learn the procedure.

You can also get RAIM predictions from DUATS or by asking for them when you contact flight service.

A final note on RAIM: As we transition to wider use of WAAS, RAIM itself will become less of a concern. WAAS receivers do not have a "receiver autonomous" RAIM function, as such. WAAS systems have other ways of detecting and alerting you to satellite problems.

But the alerts that pop up are similar. Here are sample satellite messages that you might see and hear on the WAAS-certified Chelton FlightLogic EFIS (electronic flight instrument system):

GPS FAILURE   (no GPS)
GPS INTEGRITY   (GPS loss of information)

Other Emergencies

GPS has had a remarkable spillover effect on avionics. As we have transitioned to more sophisticated and reliable space-based systems, our black-box technology in general has also become more sophisticated and reliable.

But some problems can still catch you by surprise, especially if you haven't thought about them recently. Here is a brief review of ways to handle some of the more significant non-GPS avionics problems.

Two-Way Radio Communications Failure

Time passes, and you slowly become aware that you haven't heard any radio chatter for a while. You check all your switches and frequencies. Nothing. You call the last ATC frequency you were in touch with. All quiet. You try again on 121.5—that's what it's there for, right? Still nothing.

You must now assume a two-way radio communications failure. FAR 91.185 (see Reference Section, page 209) spells out very clearly the procedures ATC will be looking for you to follow.

We are all familiar with these procedures. However, there are some crucial points that are worth emphasizing again in the following sections.

Importance of Logging Times

Note how important it is in FAR 91.185 to know the estimated time of arrival (ETA) at your destination if you have not received an "expect further clearance time." ATC will expect you to show up for an approach at an ETA based on your filed or amended estimated time en route (ETE).

The controllers at ATC know when you took off from the departure airport—do you?—and they know the estimated time en route from your filed or amended flight plan. They will add the time en route to the departure time, come up with an ETA, and reserve all approaches at the destination for this ETA plus 30 minutes. Until that time expires, you "own" that airport.

So it becomes extremely important to know what time you took off, because without that information, you'll never know when to commence the approach. ("You can't tell when you're going to get someplace if you don't know when you left someplace.")

This seems so basic, yet many pilots forget to log their takeoff time. If a designated examiner notices this on an instrument flight test, you can bet that at some point he or she will ask the applicant what is the destination ETA in the event of lost communications, or to detail the steps to take in care of communications failure.

Emergency Altitudes

A second point that always raises questions is the altitude for completing the flight in the event of a two-way radio communications failure. ATC expects the flight to continue at (1) the last assigned altitude, *or* (2) the minimum altitude for IFR operations, *or* (3) the altitude that ATC has advised you to expect in a further clearance. You will fly at the highest altitude of the three possibilities for any given leg.

You will have to make some choices. These choices hinge on how the "minimum altitude for IFR operations" is determined and whether or not it is higher than the altitude assigned or advised to expect. The most important consideration is terrain and obstacle clearance.

Along an airway, the minimum for IFR operations is the MEA (minimum en route altitude). MEA guarantees terrain and obstacle clearance.

If the en route low altitude chart shows that you are approaching a route segment with an MEA that is *higher* than the altitude assigned by ATC or what ATC advised you to expect, you would begin a climb to reach the new altitude so that you reach it prior to the point or fix where the new altitude begins. If the MEA drops down below the last assigned altitude, you would descend to the last altitude assigned by ATC or the one you were advised to expect, whichever is higher.

But suppose you were cleared direct. What is the minimum altitude *off airways* where there are no MEAs? What you need to do is determine the OROCA (off-route obstacle clearance altitude). Consult the appropriate sectional chart that covers the flight area and pick out the "maximum elevation" figure for the latitude–longitude square you occupy, then add 1,000 feet to comply with FAR minimum altitude regulations (2,000 feet in mountainous terrain; see FAR 91.177).

The maximum elevation figures are the big numbers in the center of each latitude–longitude square. You would use the maximum elevation plus 1,000 feet (or plus 2,000 feet in mountainous areas) as a substitute for the MEA. So you must carry the appropriate sectional charts with you when you are on an IFR flight plan.

The likelihood of having to use the two-way communications failure procedures is remote, but you have to know them. And "Loss of Communications" is a required task specified in the "Emergency Operations" section of the *Instrument Rating Practical Test Standards*.

In the real world of IFR you can avoid the complications of lost communications by investing in a handheld portable transceiver and carrying it on every flight. It will also be invaluable in the next scenario.

Complete Electrical Failure

Single-engine airplanes are vulnerable to complete electrical failure. There is frequently only one alternator—driven by one inexpensive V-belt—and only one voltage regulator. So, there is no backup if one of these key components fails.

The first step in the event of a complete electrical failure is to turn off all electrical equipment to minimize drain on the battery. Control the airplane by partial panel. At night you will need a flashlight to do this. Careful pilots carry two flashlights for this purpose—one as a backup in case the other starts to fade—plus extra batteries.

If the failure occurs when you are VFR, maintain VFR and land as soon as practicable.

If you are in actual IFR conditions when a complete electrical failure occurs, the procedure is equally simple, although your work is cut out for you: Find VFR conditions and land as quickly as possible.

When planning for the flight, you did locate and jot down the nearest VFR airport, right? This is where it really pays off to consult weather depiction charts for your route during the weather briefing, as mentioned before. So you know where the nearest safe haven is. If you have been logging the actual time of arrivals at each waypoint or fix along the way, you have an accurate idea of where you were when the failure occurred. Draw a straight line to the new destination from the point you intend to turn. Calculate a new magnetic heading and turn to that heading. Later, with more time, calculate an ETA at the new destination. It's like those diversions you worked so hard on during the cross-country phase of private pilot training.

When everything else is under control, set the transponder to 7700, and turn it on. Then turn on the transceiver on which you last talked to ATC. You have only a few minutes of battery power left; make sure ATC understands your predicament and intentions, and then shut everything off again. You can put the remaining battery power to good use at the VFR destination to contact tower or flash some lights.

There's a good chance that an electrical malfunction has gone undetected long enough to have already exhausted the battery. If at night, you are now flying with your flashlight in your teeth. ATC will declare an emergency on your behalf because they can't talk to you, and your transponder just dropped down to a primary target blip on their radar screen. You are on your own to get down, but ATC will move other aircraft out of the way for you. You must fly the emergency procedures so that ATC will know where you are going and what emergency altitudes they need to clear for you. Maybe you will pop into the clear and see somewhere to land.

Again, a portable transceiver will make life a lot easier. The range of these units is vastly improved if connected to an outside antenna; consider installation of an antenna jack.

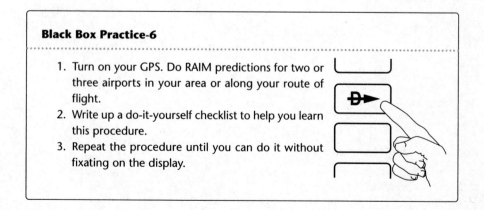

Black Box Practice-6

1. Turn on your GPS. Do RAIM predictions for two or three airports in your area or along your route of flight.
2. Write up a do-it-yourself checklist to help you learn this procedure.
3. Repeat the procedure until you can do it without fixating on the display.

9

Instrument Check Rides With GPS

To decrease the workload of FAA inspectors, the FAA decided several years ago that it was in everyone's best interest to designate the most competent flight instructors as pilot examiners. This shifted the time-consuming burden of doing practical tests from inspectors to a fee-based business conducted by these "best of the best" instructors.

Over the years, the system has worked well. Instructors being considered for designation attend training at the FAA facility in Oklahoma City. After further training, orientation, and observation by their home Flight Standards District Office (FSDO), the new examiners are given limited authorization to turn pilot applicants into certificated pilots.

Examiners are masterful instructors. For the purpose of your check ride—or practical test as the FAA calls it—the examiner is an evaluator. He or she will use exactly the same standards to test you as though you were checked by an FAA inspector. While it must be said that the examiner wants the applicant to pass the check ride, that examiner has a major responsibility to keep the airspace safe for the rest of us. Therefore, if an applicant is not a good pilot, the FAA and its team of designated examiners don't want him or her in the air.

The instrument practical test is a *rating* test as opposed to a *certification* test. Certificates are issued for the following pilot levels: recreational, private, commercial, airline transport pilot (ATP), and flight instructor. The instrument rating is an added rating for the private, commercial, and flight instructor certificates.

The examiner is not allowed to instruct during a check ride; he or she is an evaluator. However, most examiners are more than eager to give an applicant tips during the debriefing after the check ride, regardless of the outcome. Take advantage of their knowledge, advice, and wisdom.

GPS Changes Things

Practical Test Standards (PTSs) have been prepared for each certificate or rating to create consistent testing standards among examiners, inspectors, and applicants. These PTSs continue to be amended and updated for various reasons, including the growing use of GPS.

The current instrument rating PTS booklet specifies that the applicant must satisfactorily demonstrate two different types of nonprecision approaches as well as one precision ILS approach. As this was written, the following six nonprecision approaches were listed in the PTS booklet as subject to test on a check ride:

VOR   LOC
NDB   LDA
GPS   SDF

In addition, the PTS booklet further emphasizes that if you have DME, ADF, and/or GPS aboard your check ride airplane, the examiner has the option of asking you to make approaches with them. Conversely, if DME, ADF, and/or GPS are not installed aboard your plane, their absence won't affect the outcome of your flight test.

In other words, "if you have it, be prepared to use it."

Some people show up with a GPS that has an expired database and think they won't be tested on GPS approaches. Think again! If the weather is VFR, you may still be asked to do a GPS approach under VFR conditions. In this situation the designated examiner serves as safety pilot, and you will be evaluated on your ability to carry out the specified procedure.

The instrument rating standards continue to evolve to reflect the GPS revolution. For example, since you can't be tested on NDB approaches if there is no ADF aboard your aircraft, NDB approaches are beginning to fade away in importance. Many training centers are taking ADFs out of the cockpit entirely and replacing them with GPS units. We can expect more changes in the practical test standards as more and more WAAS approaches go online in coming months. It won't be long before we'll be routinely shooting WAAS nonprecision and precision approaches and being tested on them.

One of the nonprecision approaches has to be performed under partial panel conditions. Have you have been practicing approaches with your gyroscopic heading and attitude instruments covered? The partial panel requirement is another good reason to carry a GPS-equipped airplane to the check ride instead of one with ADF. With practice, the GPS approach is much simpler than the NDB approach under partial panel. GPS approaches are set up and behave like VOR or ILS approaches, and GPS is a lot less squirrely than a wobbly NDB partial-panel approach. This is especially true in a stiff crosswind.

Even if the examiner doesn't ask for an NDB approach in your ADF-equipped plane, he or she might still ask you to demonstrate the use of ADF for something else. You might have to track or hold with it, for example—especially if you have an RMI aboard. Most larger aircraft—and some smaller ones—use RMIs. The RMI provides magnetic bearings to both VORs and NDBs and is a huge help in performing DME arcs. With or without an RMI, conscientious examiners have every reason to ask you to perform some type of ADF navigation task if ADF is installed in the check ride airplane.

You Have GPS, but Is It Legal?

One of the first things a designated examiner looks at in a check ride airplane is the GPS. Is it legal for IFR use? It absolutely must be a properly and professionally installed panel unit. Yoke-mounted GPS is not approved for instrument flight. Although a handheld or yoke-mounted GPS unit may be easier to scan, the argument ends when you consider the lack of precision and reliability a temporarily mounted antenna provides. An antenna stuck to a window or an instrument panel can't provide the reception of an antenna mounted on the top of the fuselage. Any bank or change of direction is a potential for signal loss.

If you need another argument against the yoke-mounted cousin, consider how the unit is powered. If your battery-powered GPS unit starts to fade during a night approach, you are in a precarious situation. If you have a cigarette lighter plug-in, you haven't got a solid connection either. What happens when that fuse blows? Not the circuit breaker, mind you, but the fuse in the plug of the cord. Have you a handy replacement within an arm's reach? Would you need a jeweler's screwdriver to unscrew the plug? Who's flying the plane while you're fixing a cord- or battery-related problem? (Hint: not the examiner!)

Flight Planning Review

The examiner has the discretion to ask for planning with either VOR or GPS navigational systems or both. A prudent pilot (and applicant) will always back up one with the other.

Can you show up with a DUATS-generated (or other computer-generated) navigation log? The answer, in most cases, is yes. Examiners want to see a systematic approach to the assigned cross-country trip that will lead to an accurately planned and safely flown flight. But some examiners prefer a hand planned and calculated nav log. Check with your examiner before you work out your plan to see if there is any uncertainty about what he or she is looking for. Most examiners believe they can find out more about your ability to perform as an instrument pilot in this new age of technology if you show some computer savvy.

But remember: you—*and only you*—are responsible for the planning you present. If there is an error in what DUATS or some other computer program has produced for you, the examiner can't "pink slip" DUATS or your computer.

Safe Altitudes

A computer-generated flight plan/nav log might not show minimum safe altitudes. That's what the Sectional Charts are for, right? That's why we still have to say we're operating *almost* paperless. On the check ride, you will need to show that you know how to determine minimum safe altitudes (MSAs) for direct routes. Minimum en route altitudes (MEAs) shown on Enroute Charts are not applicable once you leave the Victor airways. So don't show up without the appropriate Sectional Chart(s)!

Oral Examination

Much of an applicant's knowledge can be tested during oral questioning. This can take place prior to or at any time during the flight test.

It is obviously much more practical to conduct some oral questioning on the ground before the flight test than in the air. For example, it doesn't make much sense to evaluate an applicant's knowledge of weight and balance computations during the actual flight test. However, some questioning is actually more pertinent in the air. For example, an examiner can make a much better evaluation of an applicant's knowledge of two-way radio communications failure procedures during the flight by asking, "What would you do now if you suddenly lost two-way radio communications?"

After the examiner has reviewed your FAA Form 8710-1 and other required pilot and aircraft documents (Fig. 9.1), he or she will give you several practical problems to solve. One is likely to be performance problem such as computing takeoff role and obstacle clearance distances on a high-density altitude day. You will probably be asked to work out a weight and balance problem.

Typical GPS Oral Questions

When you present your solutions to the performance and weight and balance problems and your completed flight plan/nav log, the examiner will ask you some questions to evaluate your knowledge about instrument flying, including GPS.

There are three levels of questions that might be asked as during the oral exam. First, there are questions with rote answers that all applicants should know. The second type of question is broader and evaluates the an applicant's deeper understanding of an important instrument-rating task, such as choosing an alternate. Finally, there is the third type of question based on a situation or scenario, such as "tell me how you would execute a missed approach at your destination."

If an applicant misses too many of the examiner's rote questions, there is little point in proceeding any further. Rote questions are often used as ice-breakers to help put the applicant at ease, for experience has shown that an applicant performs the

APPLICANT'S PRACTICAL TEST CHECKLIST

APPOINTMENT WITH EXAMINER:

EXAMINER'S NAME_____

LOCATION _____

DATE/TIME _____

ACCEPTABLE AIRCRAFT

- ☐ View-limiting device
- ☐ Aircraft Documents: Airworthiness Certificate
- ☐ Registration Certificate
- ☐ Rating Limitations
- ☐ Aircraft Maintenance Records: Airworthiness Inspections

PERSONAL EQUIPMENT

- ☐ Current Aeronautical Charts
- ☐ Computer and Plotter
- ☐ Flight Plan Form
- ☐ Flight Logs
- ☐ Current AIM

PERSONAL RECORDS

- ☐ Identification - Photo/Signature ID
- ☐ Pilot Certificate
- ☐ Medical Certificate
- ☐ Completed FAA Form 8710-1, Application for an Airman Certificate and/or Rating
- ☐ Airman Knowledge Test Report
- ☐ Logbook with Instructor's Endorsement
- ☐ Notice of Disapproval (if applicable)
- ☐ Approved School Graduation Certificate (if applicable)
- ☐ Examiner's Fee (if applicable)

Figure 9.1 The *Practical Test Standards* booklet for the instrument rating offers this handy checklist for applicants, plus full coverage on other pages of the tasks the examiner is required to evaluate.

whole check ride better if he or she is put at ease early in the process. If the applicant can't handle easy rote questions, the chances are that the flight test won't go well either. Here are some examples of rote questions relating to GPS that an applicant should handle easily (and bear in mind that the examiner will mix them in with non-GPS questions):

Shortcuts That Don't Shortchange-6

You can do yourself a big favor by making blank weight and balance forms ahead of time for the airplane to be used on the flight test and photocopying two or three extras. And that's not all you can do ahead of time. You can review the maintenance logs for the aircraft you will be using and mark pages with relevant inspections and Airworthiness Directive (AD) compliance with paper clips. Check to make sure the GPS manual will be aboard the flight test aircraft—it's required. Work out your flight plan/nav log a day or two before the flight test. Research the instrument approaches in the vicinity of the route assigned by the examiner. Remember the corollary to Murphy's Law: "If you don't know where something is, you're almost certain to get sent there!" Get a full weather briefing well before you meet the examiner, and be prepared to discuss any GPS NOTAMS and Temporary Flight Restrictions (TFRs) in effect along your route of flight.

You will have only 30 minutes to do weight, balance, and performance calculations; get a final weather briefing for your preplanned flight, select an alternate airport, and complete the flight plan/nav log. Some intelligent advance planning can save you considerable time and anxiety and help you to easily complete all these tasks within 30 minutes.

- How many GPS satellites are used to cover the Earth?
- How high are they above the Earth?
- What are the three basic segments that make up the GPS system?
- What is the minimum number of satellites needed to provide continuous service en route?
- How many satellites are needed to complete a GPS approach?
- On an approach chart, point out the symbol for a fly-over waypoint.
- Are you required to report a GPS outage if you encounter one?
- What kinds of data can you as a pilot get from a GPS unit?
- Is there a procedure for finding the nearest airport on this unit? What is it?
- What are the requirements for using a particular unit on a GPS approach?
- How do you know if the database is out of date?
- Does this GPS unit provide vertical guidance? What are its limitations?
- Is your aircraft flight manual required to have supplements for new procedures regarding your GPS?
- Is GPS approved as a sole means of navigation for en route flying?
- Is GPS approved as a sole means of navigation for approaches?
- What is RAIM?

Explanation-style questions require more from the applicant than just popping back answers to rote questions. Explanations require complete sentences with more detail. If an applicant begins to go off in the wrong direction, the examiner will likely probe deeper in that area and set up the question in a different way. This might be appropriate if the applicant's instructor taught a subject from a different angle.

Here are examples of GPS-related explanation questions:

- When does the sensitivity of the GPS course deviation indicator change?
- Does the GPS system provide any protection for pilots if a required satellite stops transmitting? Explain.
- Why does the GPS prompt you to reenter the altimeter setting as you approach your destination?
- What is the OBS button for?
- Explain why some waypoints on a DME overlay approach have parentheses around them.
- How does the GPS unit know what you are doing when you are flying an overlay approach where there is no final approach fix?
- Explain DGPS.
- What is WAAS?

And here are some examples of broader scenario or situational questions that you might have to handle while navigating on GPS:

- How you would handle the missed approach procedure for a specific published approach selected by the examiner?
- How can you tell if there is a problem with the satellites along your route? What should you do?
- What is the most commonly mishandled phase of a GPS approach? What steps would you take with your equipment to avoid this problem?
- ATC asks you to report "present position." What do they want and how do you find it out?
- You are in IMC immediately after takeoff, and your engine begins to run rough. How do you get back to the airport?
- ATC reports it no longer has you in radar contact. What do you need to do?

Now Let's Test Your Flying Ability!

As you and your examiner go out to the plane, the time has come to consider the whole picture of the instrument trip you are about to begin. Sure—setting up the GPS is important, but it's a part of the whole montage. As always, concentrate first on your basic preflight checks. Testing the warmth of the pitot tube takes on more meaning when you're an instrument pilot. So does making sure that the battery has ample charge and the electrical system is thoroughly tested during the run-up. Check these things as if your life depends on it.

The examiner will be watching as you set up the GPS. This will give him or her an all-important first impression of how well you know your equipment. As your hand moves around the radio stack, the examiner will get a further clue as to what will happen on the check ride. If you fumble around with the radios, you will probably fumble around with them in the air when you should be watching the instruments. Smooth movements while setting up your avionics will communicate confidence about your mastery of GPS and all its interconnections.

After all this, it will be a positive relief to get into the air!

During the flight phase you will be evaluated on your skill in copying clearances and working with ATC during IFR departure, en route, and approach phases. Your ability to fly the airplane while performing a multitude of tasks will now assume paramount importance.

The first instrument approach might be carried through to a landing from a straight-in or a circling approach. Or you might be asked to execute a missed approach, enter the missed approach holding pattern, and then obtain an IFR clearance in the air to a second airport where you will make a second type of instrument approach. Remember, the PTSs require three approaches: two nonprecision plus one precision ILS. If you have GPS aboard, you are almost certain that one of the nonprecision approaches will be some type of GPS approach.

At some point the examiner will cancel IFR (weather permitting) and do some steep turns and unusual attitudes. Count on flying some portion of the flight on partial panel, including one nonprecision approach.

Expect other simulated emergencies such as electrical failure (simulated by turning off the alternator half of the split power switch) or pitot tube icing (simulated by covering the airspeed indicator).

And yes, some examiners will conduct the flight test in actual IFR conditions. Talk about realistic! Here is how Henry Sollman, author of *Mastering Instrument Flying,* described the way he set this up when he was a designated examiner:

> When the flight plan is filed I will talk to ATC—hopefully with the controller responsible for the sector—and arrange a block of airspace in which to conduct the steep turns and unusual attitudes. I try to get a block of airspace at the assigned altitude ±1,000 feet along a 20-mile length of airway.
>
> It gets to be a little complicated for the examiner, but the applicant should not assume we won't make the flight if the weather is IFR, or that we won't do steep turns, unusual attitudes, minimum controllable airspeed, stalls, and partial panel under actual IFR conditions.

Common Deficiencies

If you bring a GPS-equipped aircraft to the check ride, you will be expected to demonstrate your mastery of GPS throughout all phases of the flight. In talking with examiners, we find that the most common GPS deficiency they see on check rides is a lack of familiarity with the GPS equipment.

This shows up as a fixation on the unit, especially on the moving map. And it is most likely to occur when changing functions and pages in flight. The symptoms of GPS fixation range from drifting off assigned altitudes and headings to complete loss of situational awareness.

Another common GPS deficiency is confusion over the GPS/NAV feed. Applicants sometimes fail to switch the navigation feed from GPS to NAV when the situation requires it, and vice versa when it's time to switch back to GPS. This is a frequent source of confusion on missed approaches if the applicant is not including the GPS/NAV toggle switch and annunciator light in his or her instrument scan. This problem can be solved by simply putting a note in the right place on the appropriate checklist.

Holding altitude consistently within the tolerances of the *Practical Test Standards* remains a problem. And it's not unusual for the examiner to encounter an occasional applicant who goes below the minimums on an instrument approach and does not take prompt corrective action.

That's why we recommend that instructors start their students working from the beginning toward the goal of remaining within 2° of heading, 2 knots of airspeed, and 20 feet in altitude. If you fly within these tolerances on an instrument flight test, it will be a big boost to your morale, and it will convincingly demonstrate to the examiner that you are a skillful instrument pilot.

NDB procedures continue to be a problem. It is not uncommon for an applicant to fly an otherwise acceptable instrument flight test, but then fail on an NDB approach. This is definitely an area that flight instructors should devote more time and attention to.

But something else is going on here when pilots cling to the NDB. Let's call it fear of GPS.

Sometimes applicants will show up for a flight test with ADF aboard but no GPS, because they have not received sufficient GPS instruction to feel competent and comfortable with it. They spend a lot of time on NDB tracking, holding, and approaches, even though GPS is fast rendering NDB obsolete. They struggle with a technology that is fading away while not preparing themselves for the future.

So here comes this applicant in a 1976 Cessna. He's successfully avoided having to demonstrate GPS during his check ride by leaving his newer, GPS-equipped plane at his home field and flying in with an ADF aboard and no GPS on this old clunker.

The examiner might have a surprise for him. If the examiner doesn't find a mechanical problem or a placard missing in this rough-looking rental plane first, he or she might begin the oral part of his practical test with about five good questions on GPS!

The greatest responsibility the examiner has is to put only safe, instrument-rated pilots in the air for the protection of those of us who are already up there. If you don't show up with a plane capable of flying GPS approaches these days, an examiner can't tell if you are up to the future challenges of precision and nonprecision approaches using GPS. And the future may be nearer than you think!

Therefore, our advice is to study this book, your PTS booklet, and any other training materials you find useful, and make sure your instructor is covering all areas that you will be tested on. We are in a transition now between ground-based navigation systems and GPS, and many instructors are not yet up to speed on the new technology. Some may even sign you off without adequate GPS instruction, and then direct you to an ancient rental Cessna 172 with no GPS aboard. Don't let it happen to you!

Black Box Practice-7

1. Turn on your GPS. Check start-up screens for correct information. Check preference pages to make sure the default choices are what you want.
2. Enter the IFR cross-country flight the examiner has assigned for the check ride—including DP, STAR, and approach. If you have access to a docking station, fly this flight a couple of times. Do a deviation around weather and a nearest diversion. Execute a missed approach with holding, and then proceed to your alternate airport.
3. If you don't have access to a docking station, enter the flight plan numbers into the GPS aboard the plane you will use on the check ride. Set up and mentally rehearse the flight you will fly as much as your equipment permits, including deviations, nearest diversions, approaches, and missed approaches.

10

Looking Ahead

It's almost unbelievable how creative thinkers are using GPS to transform our everyday cockpits. What seemed like science fiction a short time ago is now regularly advertised in the pages of our leading general aviation magazines. There always seems to be something new and breathtaking on the horizon

In this chapter, we'll discuss some of the advanced capabilities made possible by GPS as well as its accelerating possibilities for the future.

But first, just how much high tech do you need? Are you really ready to jump into a glass cockpit? Let's be practical for a few minutes.

The most basic GPS panel-mounted systems now on the market will perform all the functions we have been discussing in this book so far: entering and flying flight plans; making approaches; and coping with changes, outages, and emergencies. All the fundamentals, and much, much more. The basic system displays may be on the small side, but they nevertheless show us more at a glance than we ever had with land-based navaids.

Why not consider working with a handheld unit for a while to build up VFR experience with GPS before investing in a more sophisticated, more expensive panel-mounted system? Although handhelds are not certified for IFR use, some of the newer models coming on the market are WAAS (wide area augmentation system) capable, meaning that they receive and process the more accurate WAAS-augmented GPS signals.

Handhelds can provide excellent cross-checks for use with conventional VFR navigation. And when you move up to a panel-mounted GPS, your handheld unit can still serve as a battery-powered VFR backup if your electrical system fails completely.

Transitioning to WAAS

Eventually all GPS systems will be WAAS and/or LAAS (local area augmentation system) augmented, and we'll all be expected to make WAAS-based RNAV approaches as a matter of routine.

Are you ready for WAAS?

As we have seen in the WAAS material presented in this book, flying with WAAS is simple enough. It does its job in the background, enhancing GPS seamlessly and invisibly. RNAV approaches with WAAS are more reliable and more accurate than GPS overlays and stand-alones. WAAS LNAV/VNAV and LPV approaches add an electronic glide path to give you vertical guidance in situations where we never had it before. WAAS approaches are flown the same way you now fly VOR and ILS approaches, but with different minima.

Sooner or later WAAS will be the space-based standard—probably sooner.

How about the plane you fly most—yours or a rental? Is it ready for WAAS? What will it take to make it ready?

And what about your autopilot? Does the plane you fly most have one? Maybe thinking about GPS upgrades should start with the autopilot options now available.

Working with WAAS requires an upgrade to the basic GPS receiver. Some receivers, such as the Garmin GNS 530 and the Bendix King KLN 94, can be made WAAS-capable with a relatively inexpensive retrofit. Others, such as the Apollo CNX80 (Fig. 10.1) and the Chelton FlightLogic EFIS (Fig. 10.2) come off the shelf with WAAS already installed.

GPS Costs

Cost is probably the main issue for most of us when we think about upgrading avionics. There is a wide range of GPS equipment on the market today, as well as a substantial range of options and prices. Here are some ballpark price ranges for basic general aviation GPS receivers. (Prices, of course, are subject to change over time.)

- The handheld GPS units mentioned earlier range from $500 to $1,500 plus additional charges for optional accessories such as antenna mounts, yoke mounts, cigarette lighter power adapters, battery packs, and the like.
- Basic IFR-certified GPS receivers run from $5,000 to $12,000, depending on options and installation costs. MFD capability can added be added for about $4,000 plus installation.
- GPS with PFD and MFDs (such as the Chelton Flight Logic system) will likely run from $60,000 to $150,000, depending on configuration, options, and installation costs. This is the "glass cockpit" configuration.

Don't forget to ask how much it will cost to subscribe to an electronic database service to keep all your charts and approach plates current!

Glass Cockpits

For years we've been hearing about glass cockpits. But it took time—and the advent of GPS—for this concept to settle down and give us systems that give good results in general aviation cockpits.

Figure 10.1   The Apollo CNX80 was the first WAAS-certified general aviation GPS system. *(Courtesy of Garmin Ltd.)*

There are two problems here. One, obviously, is space. There isn't much room for cramming additional high-tech systems and displays into the typical general aviation cockpit. This problem was solved as units grew smaller, and in some newer planes, wider cockpits allowed more ample panels.

The second problem is pilot workload, and this has proven to be more of a challenge, especially in the single-pilot IFR situation.

Again and again instructors and designated examiners find that students and instrument-rating applicants continue to get confused and fall behind in their work when faced with fundamental GPS tasks such as approaches and clearance changes. Training that emphasizes the basics—good planning, cockpit familiarity, checklists, and efficient instrument scans—can make the pilot workload much less hectic and confusing. But there is another factor that must be understood and dealt with.

Display Distractions

The human eye simply cannot resist the temptation to focus on whatever is moving. Put a big, fascinating moving map with glorious color in the cockpit, and you cannot help yourself. Your eyes will be drawn to it irresistibly, just as your eyes were drawn to the vertical speed indicator (VSI) in the early days of your instrument training. Whenever that needle began moving up and down, our eyes locked on it, even though the readings weren't reliable until the needle stopped moving. If it moves, we want to watch it—it's as simple as that.

Figure 10.2 Cockpit display of the Chelton Flight Logic EFIS. *(Courtesy of Chelton Flight Systems.)*

It has taken a while, but glass cockpit configurations for general-aviation aircraft have finally evolved to the point where they actually reduce cockpit confusion. Let's take a look at what's happening and see why this is so.

The Power of the PFD

Instead of cramming display graphics into wherever room can be found for them, the newer instrument panels have been designed from scratch to group related information more logically.

HIGH-TECH ABBREVIATION AND ACRONYMS

ADC Air Data Computer
ADS-B Automatic Dependent Surveillance–Broadcast
AFM Aircraft Flight Manual
AHRS Attitude Heading Reference System
CDTI Cockpit Display of Traffic Information
EFIS Electronic Flight Instrument System
EGPWS Enhanced Ground Proximity Warning System
FMS Flight Management System
GPWS Ground Proximity Warning System
MFD Multifunction Display
PFD Primary Flight Display
RNP Required Navigation Performance
TAS Traffic Advisory System
TAWS Terrain Awareness and Warning System
TCAD Traffic Collision Alert Device
TCAS Traffic Alert and Collision Avoidance System
TIS Traffic Information Service
UAT Universal Access Transceiver
WAAS Wide Area Augmentation System

For example, all the information you need to fly the airplane and get where you want to go can now be presented in one single PFD (primary flight display) located in the center of a pilot's normal scan. See Fig. 10.3 (next page), the instrument panel of the Cirrus SR22.

At the heart of top-of-the line PFDs are advanced electronic air data, attitude, and heading sensors that work behind the scene to provide continuous streams of digital information for the PDF flight display. Gone are the old air pressure and vacuum-driven gauges that we have been using for so long. This shift to all-digital electronic inputs is one of the less noticed technical revolutions accompanying the arrival of GPS.

But don't worry! Look at Fig. 10.3 again, and you'll see a conventional grouping of airspeed indicator, attitude indicator, and altimeter near the PFD. Your old friends will still be there if you need them.

PFDs also come equipped with heading and altitude bugs for managing the autopilot, including coupled approaches if the autopilot has this advanced capability.

The Cirrus SR22 instrument panel shown in Fig. 10.3 has the PFD directly in front of the pilot, with the MFD just to the right of it. The backup conventional flight instruments are clustered just below the PFD. Twin Garmin 430 controls are within easy reach below the MFD. Engine instruments are grouped together at the far right of the panel.

Could this become a standard arrangement for the instrument panel of the future?

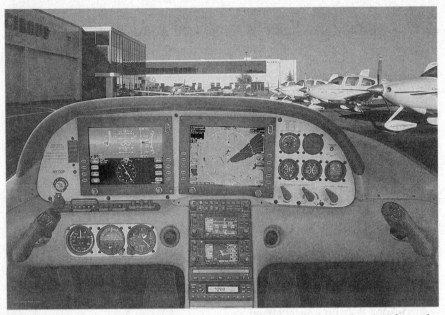

Figure 10.3 Is this the small plane cockpit of the future? This is the view from the pilot's seat of the Cirrus SR22. *(Justin Dillon.)*

What an MFD Display Can Show You

Yes, the MFD is basically a big, multicolor GPS moving map, the ultimate in situational awareness. But the MFDs now on the market can do much, much more.

In today's glass cockpits, MFDs work in partnership with PFDs to round out what a pilot needs to conduct a safe, efficient flight. There seems to be no limit to the amount or kind of information that can be called up on today's MFDs. But not all functions are available on all MFDs.

Like a computer, MFDs come bundled with a selection of basic programs, or functions, depending on the avionics package. These bundled functions tend to be basic to the management of any flight, such as terrain and special use airspace displays, engine performance readouts, and fuel status. Flight plans can be created and followed on the MFD as well as the PFD; flight plans filed on one display are automatically cross-filed on the other.

MFD Options

Beyond the basic functions that come with an MFD, there are many other options that can be added to the bundle or purchased separately.

As you consider all the options now available, it helps to develop a "menu" to assist in shaping the ultimate system you would like to have. Check out any brochures, demonstration disks, and videos that might be available. Weigh the options that will help you against the amount of cockpit attention they may require. Will your choices add to your workload or simplify it? Beware of options that require a huge amount of effort to make them work.

Let's explore some of the options now available for MFDs.

Airborne Weather Radar

Airborne weather radar has been with us for many years and has become better and better at pinpointing the location and intensity of convective activity. Radar "paints" a picture by scanning an area with high-energy radio pulses, and then measuring the strength and direction of the pulses that get reflected back. The more intense the return pulses from the heavy precipitation in a storm, the more dangerous the storm is. Radar returns in red on the display depict areas to avoid.

Chart Viewing

We're beginning to see more of a useful new MFD feature that allows you to see your aircraft positioned on moving-map, black-on-white Jeppesen charts with color highlights. Displays can be selected to show airport surface charts for taxiing and takeoff, approach plates, active flight plans, DPs, STARs, and more. Chart updates are available by subscription through Jeppesen's electronic chart service.

Could this signal the beginning of the end for the small library of paper charts that we've been fumbling around with in our cockpits all these years?

Datalink Weather

Now you can call up cockpit displays of weather maps, Nexrad coverage, and text such as TAFs and METARs. There are two approaches to providing this service.

Downlinked from Satellite

The Garmin GDL 49 downlinks the specific weather information you want from satellites in low Earth orbits. You send a request for what you want to the nearest satellite, which relays it to a ground station. The ground station puts together the package of items you requested and sends it back to the satellite. The satellite then relays your information to the Garmin receiver aboard your plane.

Uplinked from Ground Station Network

The Bendix/King KDR 510 uplinks weather information broadcast by a special nationwide network of ground stations connected to a central data source. The data broadcast by the ground stations are picked up and stored by the receiver aboard the aircraft. Specific weather items can then be called up from the onboard database and displayed whenever needed.

Lightning Detection

As the venerable Stormscope has demonstrated over the years, intense lightning activity is a good indicator of dangerous thunderstorm cells. When those lightning strikes begin popping up on your MFD screen, you can see exactly where not to go as well as where to go to get around them—if you can. Lightning detection systems are invaluable for avoiding embedded thunderstorms.

Terrain Awareness

Terrain awareness and warning systems (TAWS) were developed to help with the continuing problem of controlled flight into terrain, i.e., hitting a mountain or clipping an obstacle on a low approach. TAWS is available in Class B and Class A systems.

The Class B system is designed to meet the requirements of FAR Part 91 turbine-powered airplanes with six or more passenger seats, and for FAR Part 135 turbine-powered airplanes with six to nine seats. Class B TAWS is also known as a ground proximity warning system, or GPWS. Class B TAWS compares altimeter and GPS position inputs with a preloaded elevation database. The Class B system warns whenever the plane's altitude and course will take it too close to a database elevation.

The Class A system is designed for Part 135 operations with 10 or more seats and for all Part 121 operations. It operates the same way as Part B TAWS but is enhanced by downward and forward-looking radar inputs. Class A TAWS are also called enhanced ground proximity warning systems (EGPWS).

Traffic Advisories

"Traffic 12 o'clock! High!" That's a sample of the audible warning you get from one traffic advisory system (TAS) when another plane is getting too close. That ought to get your attention!

TAS displays show traffic in your vicinity on an MFD the way ATC controllers see it on their screens. If you have a TAS aboard, you get visible warnings on your MHD displays, as well as attention-getting audibles such as the one above.

There are two types of TAS available today, both of which use Mode S transponders to locate nearby traffic. Mode S is relatively new to general-aviation cockpits. In addition to sending out the usual four-digit code plus altitude, Mode S transponders send a unique code for each airplane whenever interrogated. Mode S also has a datalink capability. So Mode S–equipped aircraft can identify and communicate with each other as well with ground stations. There are two basic types of Mode S–based TASs: TCAS (traffic alert and collision avoidance system) and TIS (traffic information system).

TCAS (Traffic Alert and Collision Avoidance System)

TCAS is now required aboard airliners. So when a TCAS-equipped airliner sweeps the sky with its Mode S transponder, other Mode S–equipped aircraft will respond with enough information to determine if any of them are collision threats to each other.

TCAS comes in TCAS I and TCAS II versions. TCAS I provides a visual alert and leaves it up to individual pilots to work out whatever evasive maneuvers are needed. TCAS II is a more complex system. It computes evasive maneuvers based on the data being exchanged by transponders in both aircraft at risk and issues evasive commands. For example, one airplane may be told to pull up, while the other is told to dive. TCAS II is the system mandated for passenger airliners.

TIS (Traffic Information System)

TIS operates through computers on the ground at ATC locations operating with Mode S radar. The ground stations monitor and process Mode S transmissions in their local areas. (See Fig. 10.4, next page.) They send back a composite picture of all Mode S traffic they are tracking.

If you are equipped with a Mode S transponder and TIS system such as the Garmin GTX 330, you will see the same traffic picture as the ATC controllers. Your MFD will display the information you need to take evasive action. And you will get a voice warning.

Work in Progress

ADS-B (Automatic Dependent Surveillance–Broadcast)

ADS-B uses nonradar broadcast frequencies to continuously transmit and receive aircraft GPS positions and other data. This allows ATC locations and other aircraft to display traffic and data in their vicinity when complete radar coverage is not available. This new concept is being tested in Alaska, the Gulf of Mexico, and at the Embry-Riddle Aeronautical University campuses.

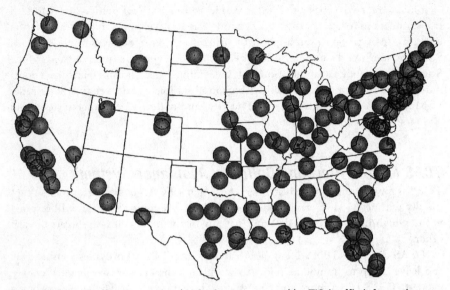

Figure 10.4 Traffic areas around U.S. airports covered by TIS (traffic information system).

For complete traffic coverage in a given area, ADS-B transceivers must be installed in all aircraft and ATC locations in that area. With ADS-B, a transceiver in your aircraft broadcasts your precise aircraft position in three-dimensional space as determined by your GPS. Likewise, your transceiver picks up data broadcast by other aircraft and displays them on your MFD or on a stand-alone screen in your cockpit called a CDTI (cockpit display of traffic information).

Air Traffic Control has access to your broadcasts—along with ADS-B broadcasts from other aircraft—through a ground station called a universal access transceiver (UAT). Through these UAT stations, ATC can identify you in places where it can't normally see you using conventional radar. Aircraft equipped with ADS-B are displayed on ATC's radar as if they were being interrogated by a radar site.

Information such as direction of flight, velocity, turns, climbs, and descents is updated every second. In Alaska, ATC routinely vectors aircraft by ADS-B to ILS final approach courses where radar coverage is not available.

Through the UAT uplink to your display, you can also receive up-to-the-minute weather, terrain, and airspace information such as temporary flight restrictions (TFRs).

Automatic dependent surveillance–broadcast can provide position information down to the surface. So an ADS-B-equipped aircraft can be identified while still taxiing on the ground. In remote locations without radar, ATC can identify ADS-B-equipped aircraft and transmit IFR clearances and departure instructions before takeoff. This could mean fewer delays to aircraft waiting to depart.

ADS-B is also effective in mountainous terrain where radar coverage is limited or nonexistent, such as the mountains and valleys of southern Alaska. The system is undergoing testing there as part of the Alaskan Capstone Project. Capstone is funded by the FAA.

Capstone's ADS-B has already helped to locate a downed pilot in Alaska, reports *AOPA Pilot* magazine. In October 2002 pilot Erick Gutierrez was flying 75 miles north from Bethel to Marshall, AK. "He didn't return, and his employer notified authorities the aircraft was overdue. In this case," says *AOPA Pilot*, "the downed aircraft's emergency locator failed to trigger."

"As a National Guard helicopter crew prepared to launch, rescue organizers turned to air traffic controllers, who called up the aircraft's ADS-B track and were able to vector a rescue helicopter directly to the aircraft's last known position. Barely two and a half hours after he was reported missing, the pilot had been picked up."

It's tempting to think that ADS-B, with its great range of capabilities, might someday replace traffic control radar. GPS makes it possible. But it would take an enormous transition to become reality.

The Big Picture

As individual pilots encountering GPS for the first time, we may not immediately need to get checked out on all the new developments. But we should know about them and open our minds to them, especially if we want to keep climbing the pilot career ladder toward a left seat in a technically advanced aircraft.

The FAA has announced its roadmap for increased aviation capacity and efficiency. By 2020 the roadmap predicts that operators will be using "RNP and RNAV procedures in all areas. A minimal operational network of ground-based navigation aids to remain in place." Will you be ready?

11

GPS FLIGHT LESSON SYLLABUS

It won't be long before GPS will be the primary air navigation system for general-aviation pilots as well as for airline transport pilots (ATPs). Right now we are still in transition, with students seeking GPS instruction at several levels.

Those seeking instrument ratings with GPS equipment aboard their check ride aircraft will be expected to demonstrate GPS proficiency, including approaches. They should complete the full syllabus of eight flight lessons and three background briefings.

For those who already have an instrument rating and wish to step up to GPS proficiency, the full syllabus might not be necessary. Flight Lesson 8 and Background Briefing 7–8 might not be needed. However, Flight Lesson 8 can serve as a GPS-based Instrument Proficiency Check (IPC) if needed.

VFR pilots without an instrument rating should be able to accomplish all they need for safe, confident GPS-based VFR operations by completing Flight Lessons 1, 2, and 3, and Background Briefing 1–2.

No set time is allotted for the eight flight lessons and the three background briefings. A lesson is complete when the completion standards are met, not when a specified amount of time has been spent on it. Instructors may repeat lessons or portions of lessons as needed for students to meet the completion standards.

The purpose of the background briefings is to introduce important new material prior to flight lessons containing new flight elements. Students should allow ample time to study the references for each background briefing on their own, dig out the answers to all questions, then write them out. *No answers are included in this syllabus.* What you learn by researching your own answers will stick with you much longer than material learned by rote!

Flight Lesson 1: Basic GPS Functions

With instructor assistance the student will plan, file, and fly an IFR GPS cross-country trip based on the [D►] function to a destination approximately 1 hour from the departure airport. He or she will make a full-stop landing and then file another IFR flight plan to return to the departure airport. The student will learn how to plan a GPS flight, prepare a flight plan/nav log, obtain weather briefings, review GPS NOTAMS, get ATC clearances, and enter the correct GPS data. The student will fly the flight under the hood, unless otherwise directed.

The instructor will demonstrate unhooded GPS approaches to full-stop landings at the destination and return airports. Ideally, the student will be able to fly most of the flight with some coaching from the instructor.

Assigned Reading

Chapter 2   Mastering GPS Flying in Eight Lessons
Chapter 3   All You Need to Know about GPS Technology
Chapter 4   Buttons, Knobs, and Switches
Chapter 5   How to Plan a GPS Flight
Chapter 6   Departure, En Route, and Arrival

Preflight Briefing

1. Introduce GPS flight planning, including weather briefings and NOTAMS.
2. Introduce filing GPS flight plans.
3. Review IFR airplane and instrument checks.
4. Introduce GPS data entry, including
   • GPS startup and opening pages
   • Use of cursors and annunciator lights
   • Entering flight plan
5. Review clearance copying.
6. Introduce GPS departure, en route, and arrival procedures, including waypoint alerting and turn anticipation.

Completion Standards

Flight Lesson 1 is complete when student has a sufficient overview of the planning, filing, and conduct of GPS IFR (or GPS VFR, if noninstrument rated) flights to begin planning and filing GPS flight plans with a minimum of assistance from the instructor. The student will enter GPS data in five minutes or less with a minimum of fixation on the GPS data display. He or she will meet instrument rating standards for holding headings and tracks within ±10°, airspeeds within ±10 knots, and altitudes within ± 100 feet.

Postflight Briefing

1. Review areas that need additional work.
2. Assign data-entry drills.
3. Go over routes and destinations assigned for Flight Lesson 2.
4. Preview new elements in Flight Lesson 2:
   - NRST (nearest) function
   - Procedures for deviating from desired track for traffic or weather and diverting to a different destination

Background Briefing 1–2: GPS Background, Data Entry, and Flight Procedures

Immediately after completing Flight Lesson 1, the student should commence work on Background Briefing 1–2, writing out as many answers as possible The instructor should plan to use this briefing to give several hours of ground instruction covering basic GPS operations.

Background Briefing 1–2 is complete when the student understands the operation of his or her GPS cockpit equipment and how the basic features apply to the planning, data entry, and conduct of an actual flight.

References

1. Departure, en route, and approach charts
2. Operating manual for the GPS system being used
3. *Aeronautical Information Manual* (AIM), sections 1-1-21, Global Positioning System (GPS), and 1-1-22, Wide Area Augmentation System (WAAS)—see Reference Section, pages 159 and 186.

Questions

1. How many satellites make up the GPS constellation?
2. Name the sources available to determine the initial desired track for a great circle route.
3. Are GPS headings referenced to true north or magnetic north?
4. Explain the procedure for switching navigational systems if the GPS unit fails en route.

5. What is RAIM, and why is it important?
6. What NOTAMS cover GPS, and how do you get them?
7. How can a pilot determine if there will be a satellite outage at his or her destination?
8. Describe the limitations on your particular unit.
9. Is it necessary to have moving-map capabilities to legally use the GPS unit?
10. Is it allowable to use a yoke-mounted GPS unit for GPS navigation in IFR flight?
11. When is active monitoring of alternate means of navigation (such as VOR) required on a GPS flight?

Flight Lesson 2: Departures and En Route GPS—Part I

With instructor assistance the student will plan, file, and fly an IFR GPS cross-country flight of approximately 2 hours with several waypoints. The student will learn how to plan a GPS flight with several waypoints, prepare a flight plan/nav log, obtain weather briefings, review GPS NOTAMS, then get ATC clearances, enter the GPS data, and fly the flight under the hood, unless otherwise directed.

The instructor will demonstrate an unhooded GPS approach to full-stop landing at the final destination. Ideally, the student will be able to fly most of the flight with some coaching from the instructor. He or she will meet instrument rating standards for holding headings and tracks within ± 10°, airspeeds within ± 10 knots, and altitudes within ± 100 feet.

Assigned Reading

Review Chapters 2 to 6 as needed.

Preflight Briefing

1. Review flight plan.
2. Review GPS data entry.
3. Review GPS departure, en route, and arrival procedures.

Completion Standards

Flight Lesson 2 is complete when the student plans, files, enters the data for a GPS flight in 30 minutes or less, and departs and transitions to the en route portion of the flight with a minimum of assistance from the instructor and with a minimum of fixation on the GPS display. He or she will maintain headings and tracks within instrument rating standards for holding headings within ±10°, airspeeds within ±10 knots, and altitudes within ±100 feet, but will begin working toward professional standards of holding heading within ± 2°, airspeed within ± 2 knots, and altitude within ± 20 feet.

Postflight Briefing

1. Review areas that need further work.
2. Assign data-entry drills.
3. Go over route, destinations, and IFR GPS approaches for Flight Lesson 3.

Flight Lesson 3: Departures and En Route GPS— Part II

The student will plan, file, obtain clearance, and depart IFR as assigned by the instructor. Approximately 20 minutes into the flight, the instructor will direct the student to cancel IFR and continue VFR to (1) practice deviating from desired track for traffic or weather, and (2) diverting to a different destination airport with an instrument approach. Flight Lesson 3 will be flown under the hood, unless otherwise directed.

The instructor will demonstrate an unhooded GPS approach to a full-stop landing at the new destination. Ideally, the student will be able to fly 60 to 90 percent of the flight with some coaching from the instructor.

Assigned Reading

Review Chapters 2–6 as needed.

Preflight Briefing

1. Review flight plan.
2. Review GPS data entry.
3. Review GPS departure, en route, and arrival procedures.
4. Introduce in-flight data-entry procedures and techniques for switching back and forth between GPS and VOR navigation.

5. Introduce procedures for deviating from desired track for traffic or weather.
6. Introduce diverting to a different destination using **⎡ Ð➤ ⎤** and NRST functions.

Completion Standards

Flight Lesson 3 is complete when the student plans, files, enters the data for a GPS flight in 30 minutes or less, and departs and transitions to the en route portion of the flight with a minimum of assistance from the instructor. The student will set up and fly deviations and diversions directed by the instructor with a minimum of fixation on the GPS display. He or she will maintain headings and tracks within professional standards of heading within ± 2°, airspeed within ±2 knots, and altitude within ± 20 feet.

Postflight Briefing

1. Review areas that need further work.
2. Assign data-entry drills.
3. Go over route, destinations, and IFR GPS approaches for Flight Lesson 4.
4. Preview new elements in Flight Lesson 4:
   • GPS stand-alone and overlay approaches
   • GPS missed approaches
   • GPS holding

Background Briefing 3–4: GPS Approaches

Immediately after completing Background Briefing 1–2, the student should commence work on Background Briefing 2–3, writing in as many answers as possible.

This briefing covers overlay and stand-alone nonprecision GPS approaches and missed approaches; holding; and common variations on full procedures such as radar vectors, and arc entries.

In addition, this background briefing introduces the student to the new GPS precision approaches with WAAS. The briefing also covers other important GPS approach situations such as RAIM and GPS outages.

Background Briefing 3–4 is complete when the student can talk through GPS overlay, stand-alone, and WAAS precision approaches, as well as missed approaches, at nearby airports.

References

1. Departure, en route, and approach charts
2. Manual for the GPS system being used

3. *Aeronautical Information Manual* (AIM), section 1-1-21, Global Positioning System (GPS), and 1-1-22, Wide Area Augmentation System (WAAS)—see Reference Section, pages 159 and 186.

Questions

1. Can you make a nonprecision GPS approach without an up-to-date database?
2. Which GPS approaches are considered nonprecision approaches? Which are considered precision approaches?
3. What is WAAS? What is LAAS? What are the similarities and the differences?
4. Are there any WAAS approaches near you?
5. Have any LAAS procedures been approved? Are there any near you?
6. Name the types of waypoints in a Basic T design GPS approach.
7. Is it possible to delete or add waypoints to the approach section of a flight plan?
8. What are the major differences between stand-alone and overlay approaches?
9. On an overlay approach, why do the waypoints have parentheses around them?
10. During an arrival at the destination, when does the approach mode change to greater sensitivity?
11. Does an alternate on a GPS flight plan have to have non-GPS approaches available? Why?
12. What do you do at the missed approach point (MAP)?
13. If the destination is below minimums, how do you proceed to the alternate using GPS?

Flight Lesson 4: GPS Instrument Approaches—Part I: Overlay and Stand-Alone Approaches

The student will fly a short cross-country flight to a nearby destination and alternate using airports with published GPS overlay and stand-alone approaches available at either airport or both. The student will fly the first approach to minimums, conduct a missed approach, fly to the alternate, fly the second type of approach there, conduct a missed approach there, and then cancel IFR and proceed VFR as directed by the instructor.

The student will then make additional GPS approaches as directed by the instructor, practicing GPS holding patterns in both the approach and the missed approach procedures whenever possible. The flight will be flown hooded, unless otherwise directed.

Assigned Reading

Chapter 7   GPS Overlay, Stand-Alone, and WAAS Approaches.
Chapter 8   Outages, Emergencies, and Other Surprises.
Review Chapters 2–6 as needed.

Preflight Briefing

1. Review GPS flight planning, filing, and data entry.
2. Review GPS departure, en route, and arrival procedures.
3. Review minimum descent altitude (MDA) and other minimum altitudes.
4. Review ceiling and visibility minimums for landing out of IFR approaches.
5. Introduce GPS overlay and stand-alone approaches and missed approaches.
6. Introduce GPS holding.

Completion Standards

Flight Lesson 4 is complete when the student is able to fly both overlay and stand-alone approaches, holding, and missed approaches with minimum assistance from the instructor and with a minimum of fixation on the GPS display. He or she will maintain headings, airspeeds, and altitudes within professional standards of "2, 2, & 20."

Postflight Briefing

1. Review areas that need further work.
2. Assign data-entry drills.
3. Assign routes and destinations for Flight Lesson 5.
4. Preview new elements in Flight Lesson 5:
   • Approach and missed approach clearance amendments, radar vectors to final approach, and missed approaches
   • Techniques for handling RAIM outages, GPS failures, and the use of backup facilities

Flight Lesson 5: GPS Approaches—Part II: Approach Amendments and Missed Approach Vectors

The student will file and fly a short cross-country flight to a nearby destination with a published GPS approach, canceling IFR at the first IAF. The instructor will then act as ATC and direct the student, using GPS, VOR, and ILS inputs, to hold, change approaches to a different runway, and comply with ATC vectors on missed approaches. The flight will be flown hooded, unless otherwise directed.

Assigned Reading

Review Chapters 2 to 8 as needed.

Preflight Briefing

1. Review MDA and other minimum altitudes.
2. Review ceiling and visibility minimums for landing out of IFR approaches.
3. Review VOR and ILS approaches.
4. Review GPS overlay and stand-alone approaches, holding, and missed approaches.
5. Review IFR lost communications procedures, loss of electrical power, and other non-GPS emergencies.
6. Review in-flight data-entry procedures and techniques for switching back and forth between GPS and VOR and/or ILS navigation equipment.
7. Introduce approach and missed approach clearance amendments, as well as radar vectors to final approach course and vectors to missed approach

Completion Standards

Flight Lesson 5 is complete when the student is able to switch back and forth between GPS and VOR/ILS navigation as required by changes in clearances, radar vectors, and simulated GPS outages with a minimum of fixation on the GPS display. He or she will maintain headings, airspeeds, and altitudes within "2, 2, & 20."

Postflight Briefing

1. Review areas that need further work.
2. Assign data-entry drills.

3. Assign route and destinations for Flight Lesson 6.
4. Preview new elements in Flight Lesson 6:
   - WAAS precision approaches
   - WAAS minimums
   - WAAS missed approaches

Flight Lesson 6: GPS Approaches—Part III: Precision Approaches with WAAS

The student will fly a short cross-country flight to a nearby destination and alternate using airports with published GPS approaches with WAAS at either or both airports. (If WAAS is not available, this Flight Lesson can be skipped.) The student will fly the first WAAS approach to minimums, conduct a missed approach, fly to the alternate, fly another approach, conduct a missed approach, and then cancel IFR and proceed VFR to practice additional approaches as directed by the instructor.

The student will practice GPS holding patterns in both the approaches and the missed approach procedures whenever possible. The flight will be flown hooded, unless otherwise directed.

Assigned Reading

Review Chapters 2 to 8 as needed.

Preflight Briefing

1. Review GPS flight planning, filing, and data entry.
2. Review GPS departure, en route, and arrival procedures.
3. Review ceiling and visibility minimums for landing out of IFR approaches.
4. Introduce minimum altitudes for a WAAS approach.
5. Introduce GPS approaches with WAAS and missed approaches from WAAS approaches.

Completion Standards

Flight Lesson 6 is complete when the student is able to fly WAAS approaches, plus holding and missed approaches with minimum assistance from the instructor and with a minimum of fixation on the GPS display. He or she will maintain headings, airspeeds, and altitudes within professional standards of "2, 2, & 20."

Postflight Briefing

1. Review areas that need further work.
2. Assign data-entry drills.
3. Assign routes and destinations for Flight Lesson 7.
4. Review GPS cross-country communications and navigation procedures, including void time clearances.

Flight Lesson 7: Long IFR Cross-Country with GPS

This flight can be used to satisfy the requirements of FAR 61.65 (d) (iii) for one 250-mile cross-country in simulated or actual IFR conditions, using GPS on federal airways, or as routed by ATC, including three different kinds of instrument approaches, at least one of which will utilize GPS.

The student will file and fly separate IFR GPS flight plans for each leg of the 250-mile cross-country. Each approach will be made to a full-stop landing, and the student will refile an IFR flight plan after each full-stop landing. At least one approach and landing will be made to an uncontrolled airport; at least one other approach and landing will be made to a tower-controlled airport. The flight will be flown hooded unless otherwise directed.

Assigned Reading

1. Review Chapter 2 as needed.
2. Review *Aeronautical Information Manual* (AIM), sections 1-1-21, Global Positioning System (GPS), and 1-1-22, Wide Area Augmentation System (WAAS)—see Reference Section, pages 159 and 186.

Preflight Briefing

1. Review planning and filing IFR GPS flight plans.
2. Review GPS departure, en route, and arrival procedures.
3. Review GPS nonprecision and GPS WAAS approaches.
4. Review data-entry techniques.
5. Review ILS and VOR approach procedures.

Completion Standards

Flight Lesson 7 is complete when the student demonstrates competence in organizing, planning, and flying a long, complex cross-country with a variety of approaches and destinations. He or she must also demonstrate competence in switching back and forth between GPS and VOR navigation as required by changes in clearances, radar vectors, and simulated GPS outages with a minimum of fixation on the GPS display. He or she will maintain headings, airspeeds, and altitudes within "2, 2, & 20."

Postflight Briefing

1. Review areas that require further work.
2. Preview how Flight Lesson 8 will be conducted.

Background Briefing 7–8: Flight Checks with GPS

Immediately after completing Background Briefing 2–3, the student should begin preparing Background Briefing 7–8. The emphasis in this final background briefing is on typical questions a designated examiner might ask the student to demonstrate during a flight test. The examiner's questions may relate to basic theory and operation of GPS, as well as to basic GPS equipment features and how they relate to the conduct of a check ride.

Background Briefing 7–8 is complete when the student can promptly and accurately answer an examiner's questions.

Assigned Reading

Chapter 9   Instrument Check Rides with GPS.
Review Chapters 2 to 8 as needed.

References

1. *Instrument Rating Practical Test Standards*, FAA-S-8081-4C
2. Departure, en route, and approach charts
3. Manual for the equipment being used
4. *Aeronautical Information Manual* (AIM), sections 1-1-21, Global Positioning System (GPS), and 1-1-22, Wide Area Augmentation System (WAAS)—see Reference Section, pages 159 and 186.

Questions

1. When is a GPS approach required by a designated examiner on an instrument rating check flight?
2. What are the differences among LNAV, RNAV, and VNAV?
3. Are distances on your GPS equivalent to distances determined by DME?
4. What is the difference between a fly-over waypoint and a fly-by waypoint?
5. Explain to–to navigation and waypoint sequencing.
6. Where does waypoint sequencing begin and end?
7. What are the shortcomings of turning anticipation?
8. Using a local approach plate, explain when to descend during the approach.
9. At what point during a flight should the pilot verify that the GPS/NAV button is switched to NAV? (*Hint*: a switch may be needed at several different points.)
10. Does your GPS unit automatically switch to localizer when a localizer frequency is selected?

Flight Lesson 8: Preparation for the GPS-Based FAA Instrument Check Ride

This flight is a dress rehearsal for an FAA check ride in which the instructor will play the role of the FAA examiner. Flight Lesson 8 can also serve as an Instrument Proficiency Check (IPC), if one is needed. The flight will be hooded, unless otherwise directed.

The instructor will direct the student to plan a short IFR cross-country that will utilize GPS as the primary means of navigation, with VOR and ILS as a backup. The student will carry out all the appropriate tasks specified in the *Instrument Rating Practical Test Standards*, FAAS-8081-4C.

Assigned Reading

1. Review *Instrument Rating Practical Test Standards*.
2. Review local GPS, ILS, and VOR approach procedures.

Preflight Briefing

The instructor will assign a cross-country flight to be preplanned and brought to the check ride rehearsal flight. The student will obtain a weather briefing, carry out other planning as directed (such as a weight and balance calculation), and then file the flight and proceed as filed until the instructor asks for changes.

Completion Standards

Flight Lesson 8 is complete when the student plans and files the flight in 30 minutes or less and then enters the appropriate GPS data in 5 minutes or less. The flight must be conducted so that the outcome of every maneuver and procedure is never in doubt, and no minimums are violated. He or she will maintain headings, airspeeds, and altitudes within "2, 2, & 20."

12

REFERENCE SECTION

I. Sources and Resources

Aviation Supplies & Academics (ASA)—information, sales, support
www.asa2fly.com
1-800-ASA 2FLY
Aviation Supplies & Academics, Inc.
7005 132nd Place SE
Newcastle, WA 98059-3153

Bendix/King—information, sales, support
www.bendixking.com
913-712-0400
Honeywell International Inc.
23500 West 105th Street
Olathe, KS 66061

Federal Aviation Administration (FAA)—headquarters and general
information source
www1.faa.gov
1-800-FAA-SURE
Federal Aviation Administration
800 Independence Avenue, SW
Washington, DC 20591

Garmin—information, sales, support
www.garmin.com
913-397-8200
Garmin International, Inc.

1200 East 151st Street
Olathe, KS 66062-3426

GPS Product Teams—FAA coordinating body for satellite navigation developments
http://gps.faa.gov
800 Independence Avenue, SW
Washington, DC 20591

Interagency GPS Executive Board—official GPS governing board
www.igeb.gov
202-482-5809
IGEB Executive Secretariat
4800B Herbert C. Hoover Building
Washington, DC 20230

Lone Star—docking stations
www.lonestaraviation.com
682-518-8882 or 817-633-6103
Lone Star Aviation Corp.
604 South Wisteria Street
Mansfield, TX 76063-2423

Sporty's Pilot Shop—charts, publications, videos, and accessories
www.sportys.com
1-800-SPORTYS (1-800-776-7897)
Sporty's Pilot Shop
Clermont County/Sporty's Airport
Batavia, OH 45102-9747

UPS Apollo—information, sales, support
www.garmin.com

II. Selective Availability

Official statement by the FAA concerning the end of "selective availability" and the establishment of the current level of GPS accuracy:

Conceived in the 1970s, the Global Positioning System (GPS) was originally built for military use. GPS remained a military-only technology until the early 1980s, when President Reagan declared the technology could be adapted for public use as well. By the early 1990s civilians could buy GPS equipment that was accurate within only about 300 feet. This inaccuracy was due to the deliberate distortion of the signal in order to prevent civilian gear from being used in a military attack on the U.S. This was called Selective Availability (SA).

On May 1, 2000, President Clinton signed an order ending SA as part of an ongoing effort to make GPS more attractive to civil and commercial users worldwide. Now, GPS is accurate within 40 feet, or much better. Military GPS is even more precise and has a margin of error of only a few centimeters.

The end of Selective Availability was a major turning point that has helped GPS become a global utility, now being used around the world in many different applications.

After the attacks of September 11th, the industry buzzed over the possibility of a return to SA. However, on Sept. 17, 2001, the Interagency GPS Executive Board (IGEB), which governs the GPS system, announced the United States has no intent to ever use Selective Availability again.

GPS Basics: GPS Policy-Selective Availability
(FAA briefing presentation)

III. DUATS Details

DUATS (Direct User Access Terminal System) is a free computer service for all pilots with current medical certificates, as well as others without current medicals in some excepted categories. To use DUATS you must first register with one (or both) services that provide it. DUATS is available by computer in the contiguous 48 states and is delivered by two companies, CSC (formerly DynCorp) DUATS and DTC DUAT.

1. **CSC DUATS**—provides free, immediate online access to FAA-approved services and information, including
   - Current, continuously updated aviation weather
   - Plain language weather printouts
   - Automated flight planning, including planning for GPS flights
   - Flight plan filling and closing
   - Direct dial access and enhanced graphics through CSC DUATS's Cirrus program CSC DUATS contact numbers are
     - www.duats.com—for registration and help, downloads, and information 24 hours per day, 7 days per week.
     - 1-800- 345-3828—for registration, help, and ordering free Cirrus CD ROMs 24 hours per day, 7 days per week.
     - 1-800-767-9989—direct, telenet-style terminal interface 24 hours per day, 7 days per week. (This is also the direct dial number used to access the Cirrus program.)

2. **DTC DUAT**—is provided by Data Transformation Corp and offers the
same range of services as CSC DUATS. DTC's DUAT Windows 3.04 is
comparable to DynCorp's Cirrus program and is likewise available free of
charge on CD ROM. Here are the DTC DUAT contact numbers:
- www.duat.com—registration, help, downloads, and information 24
  hours per day, 7 days per week. (Be careful! The DTC Internet address
  has no "S" at the end of DUAT.)
- 1-800-243-3828—for registration, help, and free DUAT Windows 3.04
  CD ROMs 24 hours per day, 7 days per week.
- 1-800-245-3838—direct, telenet-style terminal interface 24 hours per
  day, 7 days per week. (This is also the direct dial number used to access
  DUAT Windows 3.04.)

IV. Abbreviations and Acronyms

The arrival of GPS has brought with it more "alphabet soup" than any other aviation development in recent years. Here is a compilation of common terms used by
government and industry when referring to GPS and other new avionic technology.

A

AFCS Automatic Flight Control System

ADC Air Data Computer

ADS-B Automatic Dependent Surveillance–Broadcast

AHRS Altitude and Heading Reference System

AFM Aircraft Flight Manual

AHRS Attitude-Heading Reference System

ANP Actual Navigation Performance

APV Approach with Vertical Guidance

ATE Along Track (straight-line) Distance

B

BBS Bulletin Board System

C

CDTI   Cockpit Display of Traffic Information

CPDLC   Controller Pilot Data Link Communications

CRM   Cockpit Resource Management

D

DA   Decision Altitude

DFCS   Digital Flight Control System

DP   Departure Procedure

DUATS   Direct User Access Terminal System

DTK   Desired Track

E

ECS   Environmental Control System

EFIS   Electronic Flight Instrument System

EGPWS   Enhanced Ground Proximity Warning System

ETA   Estimated Time of Arrival

ETE   Estimated Time En route

EIDS   Electronic Instrument Display System

EVS   Enhanced Vision System

F

FAWP   Final Approach Waypoint

FB   Fly-By

FD   Flight Director System

FISDL   Flight Information Services Data Link

FMS   Flight Management System

FMSP Flight Management System Procedure

FO Fly-Over

G

GBAS Ground-Based Augmentation System

GEO Geostationary Earth Orbiting satellite

GLS Global navigation system Landing System

GNSS Global Navigation Satellite System

GNSSP Global Navigation Satellite System Panel

GPS Global Positioning System

GPWS Ground Proximity Warning System

GUS Ground Uplink Station

H

HITS Highway in the Sky

I

IAWP Initial Approach Waypoint

IAF Initial Approach Fix

ICAO International Civil Aviation Organization

IGEB Interagency GPS Executive Board

IMC Instrument Meteorological Conditions

INS Inertial Navigation System

J-K-L

LAAS Local Area Augmentation System

LNAV Lateral Navigation

LOP Line-of-Position

LORAN Long Range Navigation System

M

MAHWP Missed Approach Holding Waypoint

MAP Missed Approach Point

MSA Minimum Safe Altitude

MAWP Missed Approach Waypoint

MDA Minimum Descent Altitude

MFD Multifunction Display

MTI Moving Target Indicator

N

NAS National Airspace System

NEXRAD Next Generation Weather Radar

NPA Nonprecision Approach

O

ODP Obstacle Departure Procedure

OROCA Off-Route Obstruction Clearance Altitude

ORS Operational Revision Status

P

PA Precision Approach

PAPI Precision Approach Path Indicator

PPS Precise Positioning Service

PFD Primary Flight Display

Q-R

RAAS Runway Awareness and Advisory System

RAIM Receiver Autonomous Integrity Monitoring

RNAV Area Navigation

RNP Required Navigation Performance

S

SA Selective Availability

SBAS Satellite-Based Augmentation System

SCAT-1 DGPS Special Category I Differential GPS

SPS Standard Positioning Service

STAR Standard Terminal Arrival

SUA Special Use Airspace

T

TAS Traffic Advisory System (also True Airspeed)

TAWS Terrain Awareness and Warning System

TCAD Traffic Collision Alert Device

TCAS Traffic Alert and Collision Avoidance System

TIS Traffic Information System

TDWR Terminal Doppler Weather Radar

TFR Temporary Flight Restriction

TIS Traffic Information Service

TK Track

U

UAT Universal Access Transceiver

UTC Coordinated Universal Time

V

VDA   Vertical Descent Angle

VDP   Visual Descent Point

VIP   Video Integrator Processor

VMC   Visual Meteorological Conditions

VNAV   Vertical Navigation

VTF   Vectors to Final

W

WAAS   Wide Area Augmentation System

WGS-84   World Geodetic System of 1984

WMS   Wide-Area Master Station

WP   Waypoint

WRS   Wide-area ground Reference Station

V. Glossary

GPS-related excerpts from the Pilot/Controller Glossary in the *Aeronautical Information Manual* (AIM), *Advisory Circular* AC 20-138, Airworthiness Approval of Global Positioning System (GPS) Navigation for Use as a VFR and IFR Supplemental Navigation System, and other official sources.

A

along track distance (LTD)   The distance measured from a point-in-space by systems using area navigation reference capabilities that are not subject to slant range errors.

area navigation (RNAV)   Area navigation provides enhanced navigational capability to the pilot. RNAV equipment can compute the airplane position, actual track, and ground speed and then provide meaningful information relative to a route of flight selected by the pilot. Typical equipment will provide the pilot with distance, time, bearing, and cross-track error relative to the selected "TO" or "active" waypoint and the selected route. Several distinctly different navigational systems with different navigational performance characteristics are capable of providing area navigational functions. Present day RNAV includes INS, LORAN, VOR/DME, and GPS systems. Modern multisensor systems can integrate one or more of these systems to provide a more accurate and reliable navigational system. Because of the different levels of performance, area navigational capabilities can satisfy different levels of required navigational performance (RNP). The major types of equipment are

a. VORTAC referenced or course line computer (CLC) systems, which account for the greatest number of RNAV units in use. To function, the CLC must be within the service range of a VORTAC.

b. OMEGA/VLF, although two separate systems, can be considered as one operationally. A long-range navigation system based upon very-low-frequency radio signals transmitted from a total of 17 stations worldwide.

c. Inertial (INS) systems, which are totally self-contained and require no information from external references. They provide aircraft position and navigation information in response to signals resulting from inertial effects on components within the system.

d. MLS area navigation (MLS/RNAV), which provides area navigation with reference to an MLS ground facility.

e. LORAN-C is a long-range radio navigation system that uses ground waves transmitted at low frequency to provide user position information at ranges of up to 600 to 1,200 nautical miles at both en route and approach altitudes. The usable signal coverage areas are determined by the signal-to-noise ratio, the envelope-to-cycle difference, and the geometric relationship between the positions of the user and the transmitting stations.

f. GPS is a space-base radio positioning, navigation, and time-transfer system. The system provides highly accurate position and velocity information, and precise time, on a continuous global basis, to an unlimited number of properly equipped users. The system is unaffected by weather, and provides a worldwide common grid reference system.

area navigation (RNAV) approach configuration

a. *Standard T.* An RNAV approach whose design allows direct flight to any one of three initial approach fixes (IAF) and eliminates the need for procedure turns. The standard design is to align the procedure on the extended centerline with the missed approach point (MAP) at the runway threshold, the final approach fix (FAF), and the initial approach/intermediate fix (IAF/IF). The other two IAFs will be established perpendicular to the IF.

b. *Modified T.* An RNAV approach design for single or multiple runways where terrain or operational constraints do not allow for the standard T. The "T" may be modified by increasing or decreasing the angle from the corner IAF(s) to the IF or by eliminating one or both corner IAFs.

c. *Standard I.* An RNAV approach design for a single runway with both corner IAFs eliminated. Course reversal or radar vectoring may be required at busy terminals with multiple runways.

d. *Terminal arrival area (TAA).* The TAA is controlled airspace established in conjunction with the standard or modified T and I RNAV approach configurations. In the standard TAA, there are three areas: straight-in, left base, and right base. The arc boundaries of the three areas of the TAA are published portions of the approach and allow aircraft to transition from the en route structure direct to the nearest IAF. TAAs will also eliminate or reduce feeder routes, departure extensions, and procedure turns or course reversal.

1. *Straight-in area.* A 30-nm arc centered on the IF bounded by a straight line extending through the IF perpendicular to the intermediate course.

2. *Left-base area.* A 30-nm arc centered on the right corner IAF. The area shares a boundary with the straight-in area, except that it extends out for 30 nm from the IAF and is bounded on the other side by a line extending from the IF through the FAF to the arc.

3. *Right-base area.* A 30-nm arc centered on the left corner IAF. The area shares a boundary with the straight-in area, except that it extends out for 30 nm from the IAF and is bounded on the other side by a line extending from the IF through the FAF to the arc.

availability   The availability of the GPS system is the percentage of time that the services of the GPS system are usable. Availability is an indication of the ability of the system to provide usable service within the specified coverage area. Signal availability is the percentage of time that navigational signals transmitted from the satellites are available for use.

B

baro-aiding   Supplying pressure altitude to the onboard GPS computer; baro, for short. The baro is set as part of the GPS preflight checks.

barometric altitude   Altitude in the Earth's atmosphere above mean standard sea level pressure datum, measured by a pressure (barometric) altimeter and corrected for local barometric pressure setting.

C

controller-pilot data link communications (CPDLC)   A two-way digital very-high-frequency (VHF) air–ground communications system that conveys textual air traffic control messages between controllers and pilots.

course

a. The intended direction of flight in the horizontal plane measured in degrees from north.

b. The ILS localizer signal pattern usually specified as the front course or the back course.

c. The intended track along a straight, curved, or segmented MLS path. (See also **track**.)

D

desired track The planned or intended track between two waypoints. It is measured in degrees from either magnetic or true north. The instantaneous angle may change from point to point along the great circle track between waypoints.

differential GPS A technique used to improve GPS system accuracy by determining positioning error from the GPS satellites at a known fixed location and subsequently transmitting the determined error, or corrective factors, to GPS users operating in the same area.

E

en route operations The phase of navigation covering operations between departure and arrival terminal phases. The en route phase of navigation has two subcategories: en route domestic and en route oceanic and remote.

en route domestic The phase of flight between departure and arrival terminal phases, with departure and arrival points within the U.S. National Airspace System (NAS).

en route oceanic and remote The phase of flight between the departure and arrival terminal phases, with an extended flight path over an ocean.

estimated time en route (ETE) The estimated flying time from departure point to destination (lift-off to touchdown).

estimated time of arrival (ETA) The time the flight is estimated to arrive at the gate (scheduled operators) or the actual runway on times for nonscheduled operators.

F

flight management system A computer system that uses a large database to allow routes to be preprogrammed and fed into the system by means of a data loader. The system is constantly updated with respect to position accuracy by reference to conventional navigation aids. The sophisticated program and its associated database ensures that the most appropriate aids are automatically selected during the information update.

flight technical error (FTE)   Navigation error introduced by the pilot's (or autopilot's) ability to use displayed guidance information to track the desired flight path.

fly-by waypoint   A fly-by waypoint requires the use of turn anticipation to avoid overshoot of the next flight segment.

fly-over waypoint   A fly-over waypoint precludes any turn until the waypoint is overflown and is followed by an intercept maneuver of the next flight segment.

G

geometric altitude   Altitude above the surface of the WGS-84 ellipsoid.

global navigation system landing system (GLS)   Must have WAAS (wide area augmentation system) equipment approved for precise approach. If the GLS minimum line does not contain "PA," then the runway environment does not support precision requirements.

global positioning system (GPS)   A space–base radio positioning, navigation, and time-transfer system. The system provides highly accurate position and velocity information, and precise time, on a continuous global basis, to an unlimited number of properly equipped users. The system is unaffected by weather and provides a worldwide common grid reference system. The GPS concept is predicated upon accurate and continuous knowledge of the spatial position of each satellite in the system with respect to time and distance from a transmitting satellite to the user. The GPS receiver automatically selects appropriate signals from the satellites in view and translates these into three-dimensional position, velocity, and time. System accuracy for civil users is normally 100 meters horizontally.

global positioning system (GPS) equipment classes A(), B(), and C()
GPS equipment is categorized into the following classes (ref. TSO-C129):
a. *Class A()*. Equipment incorporating both the GPS sensor and navigation capability. This equipment incorporates receiver autonomous integrity monitoring (RAIM). Class A1 equipment includes en route, terminal, and nonprecision approach navigation capability. Class A2 equipment includes en route and terminal navigation capability only.
b. *Class B()*   Equipment consisting of a GPS sensor that provides data to an integrated navigation system (i.e., flight management system, multisensor navigation system, etc.). Class B1 equipment includes RAIM and provides

en route, terminal, and nonprecision approach capability. Class B2 equipment includes RAIM and provides en route and terminal capability only. Class B3 equipment requires the integrated navigation system to provide a level of GPS integrity equivalent to RAIM and provides en route, terminal, and nonprecision approach capability. Class B4 equipment requires the integrated navigation system to provide a level of GPS integrity equivalent to RAIM and provides en route and terminal capability only.

c. *Class C()* Equipment consisting of a GPS sensor that provides data to an integrated navigation system (i.e., flight management system, multisensor navigation system, etc.), which provides enhanced guidance to an autopilot or flight director in order to reduce flight technical error. Installation of Class C() equipment is limited to aircraft approved under 14 CFR part 121 or equivalent criteria. Class C1 equipment includes RAIM and provides en route, terminal, and nonprecision approach capability. Class C() equipment includes RAIM and provides en route and terminal capability only. Class C3 equipment needs the integrated navigation system to provide a level of GPS integrity equivalent to RAIM and provides en route, terminal, and nonprecision approach capability. Class C4 equipment requires the integrated navigation system to provide a level of GPS integrity equivalent to RAIM and provides en route and terminal capability only.

H

horizontal dilution of precision (HDOP)   A measure of the satellite geometric effects that degrade a user's horizontal position determination.

I

IF/IAWP   Intermediate fix/initial approach waypoint. The waypoint where the final approach course of a T approach meets the crossbar of the T. When designated (in conjunction with a TAA) this waypoint will be used as an IAWP when approaching the airport from certain directions and as an IFWP when beginning the approach from another IAWP.

IFWP   Intermediate fix waypoint.

initial approach fix (IAF)   The fixes depicted on instrument approach procedure charts that identify the beginning of the initial approach segment(s).

instrument approach procedure (IAP)   A series of predetermined maneuvers for the orderly transfer of an aircraft under instrument flight conditions from the beginning of the initial approach to a landing or to a point from which a landing may be made visually. It is prescribed and approved for a specific airport by competent authority.

a. U.S. civil standard instrument approach procedures are approved by the FAA as prescribed under Part 97 and are available for public use.

b. U.S. military standard instrument approach procedures are approved and published by the Department of Defense.

c. Special instrument approach procedures are approved by the FAA for individual operators but are not published in Part 97 for public use.

instrument departure procedure (DP)   A preplanned instrument flight rule (IFR) air traffic control departure procedure printed for pilot use in graphic and/or textual form. DPs provide transition from the terminal to the appropriate en route structure.

instrument meteorological conditions (IMC)   Meteorological conditions expressed in terms of visibility, distance from cloud, and ceiling less than the minima specified for visual meteorological conditions. [See also **visual meteorological conditions (VMC)**].

integrity   The probability that the system will provide accurate navigation as specified or timely warnings to users when GPS data should not be used for navigation.

intermediate fix (IF)   The fix that identifies the beginning of the intermediate approach segment of an instrument approach procedure. The fix is not normally identified on the instrument approach chart as an intermediate fix.

J

jamming   Electronic or mechanical interference that may disrupt the display of aircraft on radar or the transmission or reception of radio communications or navigation.

K-L

lateral navigation (LNAV)   A function of area navigation (RNAV) equipment that calculates, displays, and provides lateral guidance to a profile or path. [See also **vertical navigation (VNAV)**].

M

mask angle   A fixed elevation angle referenced to the user's horizon below which GPS satellites are ignored by the receiver software.

minimum descent altitude   The lowest altitude, expressed in feet above mean sea level, to which descent is authorized on final approach or circling-to-land maneuvering in execution of a standard instrument approach procedure where no electronic glideslope is provided.

missed approach

a. A maneuver conducted by a pilot when an instrument approach cannot be completed to a landing. The route of flight and altitude are shown on instrument approach procedure charts. A pilot executing a missed approach prior to the missed approach point (MAP) must continue along the final approach to the MAP. The pilot may climb immediately to the altitude specified in the missed approach procedure.

b. A term used by the pilot to inform ATC that he is executing the missed approach.

c. At locations where ATC radar service is provided, the pilot should conform to radar vectors when provided by ATC in lieu of the published missed approach procedure.

missed approach point   A point prescribed in each instrument approach procedure at which a missed approach procedure shall be executed if the required visual reference does not exist.

N

nonprecision approach operations   Those flight phases conducted on charted instrument approach procedures (lAPs) commencing at the initial approach fix and concluding at the missed approach point or the missed approach holding fix, as appropriate.

O-P

pressure altitude   Altitude in the Earth's atmosphere above mean standard sea level pressure datum plane, measured by a pressure (barometric) altimeter set to standard pressure (29.92 inches of mercury).

pseudo-range   The distance from the user to a satellite plus an unknown user clock offset distance. With four satellite signals it is possible to compute position and offset distance.

Q-R

receiver autonomous integrity monitoring (RAIM) A technique whereby a civil GPS receiver or processor determines the integrity of the GPS navigation signals using only GPS signals or GPS signals augmented with altitude. This determination is achieved by a consistency check among redundant pseudo-range measurements. At least one satellite in addition to those required for navigation must be in view for the receiver to perform the RAIM function.

required navigation performance A statement of the navigational performance necessary for operation within a defined airspace. The following terms are commonly associated with RNP:

a. Required navigation performance level or type (RNP-X). A value, in nautical miles (nm), from the intended horizontal position within which an aircraft would be at least 95 percent of the total flying time.

b. Required navigation performance (RNP) airspace. A generic term designating airspace, route(s), leg(s), operation(s), or procedure(s) where minimum required navigational performance (RNP) has been established.

c. Actual navigation performance (ANP). A measure of the current estimated navigational performance. Also referred to as estimated position error (EPE).

d. Estimated position error (EPE). A measure of the current estimated navigational performance. Also referred to as actual navigation performance (ANP).

e. Lateral navigation (LNAV). A function of area navigation (RNAV) equipment which calculates, displays, and provides lateral guidance to a profile or path.

f. Vertical navigation (VNAV). A function of area navigation (RNAV) equipment which calculates, displays, and provides vertical guidance to a profile or path.

RNAV See **area navigation**.

S

stand-alone GPS navigation system Stand-alone GPS equipment is equipment that is not combined with other navigation sensors or navigation systems such as DME, Loran C, Omega, Inertial, and the like. Stand-alone GPS equipment can, however, include other augmentation features such as altimetry smoothing, clock coasting, and so forth.

standard terminal arrival (STAR) A preplanned instrument flight rule (IFR) air traffic control arrival procedure published for pilot use in graphic or textual form. STARs provide transition from the en route structure to an outer fix or an instrument approach fix or arrival waypoint in the terminal area.

supplemental air navigation system An FAA-approved navigation system that can be used for navigation provided that an alternate navigation system, which meets all of the regulatory requirements for the route of flight, is also installed on the aircraft.

system availability   The percentage of time (specified as 98 percent) that at least 21 of the 24 GPS satellites must be operational and providing a usable navigation signal.

T

terminal area operations   Those flight phases conducted on charted standard instrument departures (SIDs), on standard terminal arrivals (STARs), or other flight operations between the last en route fix or waypoint and an initial approach fix or waypoint.

track   The actual flight path of an aircraft over the surface of the Earth. (See also **course.**)

track angle error   Track angle error is the difference between the desired track and actual track (magnetic or true).

traffic alert and collision avoidance system (TCAS)   An airborne collision-avoidance system based on radar beacon signals that operates independent of ground-based equipment. TCAS-I generates traffic advisories only. TCAS-II generates traffic advisories, and resolution (collision avoidance) in the vertical plane.

transition
a. The general term that describes the change from one phase of flight or flight condition to another, for example, transition from en route flight to the approach or transition from instrument flight to visual flight.
b. A published procedure (DP transition) used to connect the basic DP to one of several en route airways or jet routes, or a published procedure (STAR transition) used to connect one of several en route airways or jet routes to the basic STAR.

U

user-defined waypoints   Customized waypoints that can be set up in a GPS system to locate private airports and other frequently used locations.

V

vertical navigation (VNAV)   A function of area navigation (RNAV) equipment that calculates, displays, and provides vertical guidance to a profile or path. [See also **lateral navigation (LNAV).**]

visual descent point   A defined point on the final approach course of a non-precision straight-in approach from which normal descent from the MDA to the runway touchdown point may be commenced, provided the approach threshold of that runway, or approach lights, or other markings identifiable with the approach end of that runway are clearly visible to the pilot.

visual meteorological conditions (VMC)   meteorological conditions expressed in terms of visibility, distance from cloud, and ceiling equal to or better than specified minima. [See also **instrument meteorological conditions (IMC).**]

W-X-Y-Z

waypoint   A predetermined geographical position used for route and instrument approach definition, progress reports, visual reporting points, or points for transitioning or circumnavigating controlled or special-use airspace. Waypoints are defined relative to a VORTAC station or in terms of latitude/longitude coordinates. (See also **user-defined waypoints.**)

wide area augmentation system (WAAS)   The WAAS is a satellite navigation system consisting of the equipment and software which augments the GPS standard positioning service (SPS). The WAAS provides enhanced integrity, accuracy, availability, and continuity over and above GPS SPS. The differential connection function provides improved accuracy required for precision approach.

VI. AIM Excerpts Relating to GPS

Note: *These AIM excerpts contain changes through August 7, 2003. The FAA updates and publishes new AIM editions twice a year, in February and August.*

1-1-20. Global Positioning System (GPS)

a. System Overview
1. GPS is a U.S. satellite-based radio navigational, positioning, and time transfer system operated by the Department of Defense (DOD). The system provides highly accurate position and velocity information and precise time on a continuous global basis to an unlimited number of properly-equipped users. The system is unaffected by weather and provides a worldwide common grid reference

system based on the Earth-fixed coordinate system. For its Earth model, GPS uses the World Geodetic System of 1984 (WGS-84) datum.

2. GPS provides two levels of service: Standard Positioning Service (SPS) and Precise Positioning Service (PPS). SPS provides, to all users, horizontal positioning accuracy of 100 meters, or less, with a probability of 95 percent and 300 meters with a probability of 99.99 percent. PPS is more accurate than SPS; however, this is limited to authorized U.S. and allied military, federal government, and civil users who can satisfy specific U.S. requirements.

3. GPS operation is based on the concept of ranging and triangulation from a group of satellites in space which act as precise reference points. A GPS receiver measures distance from a satellite using the travel time of a radio signal. Each satellite transmits a specific code, called a coarse acquisition (C/A) code, which contains information on the satellite's position, the GPS system time, and the health and accuracy of the transmitted data. Knowing the speed at which the signal traveled (approximately 186,000 miles per second) and the exact broadcast time, the distance traveled by the signal can be computed from the arrival time.

4. The GPS receiver matches each satellite's C/A code with an identical copy of the code contained in the receiver's database. By shifting its copy of the satellite's code in a matching process, and by comparing this shift with its internal clock, the receiver can calculate how long it took the signal to travel from the satellite to the receiver. The distance derived from this method of computing distance is called a pseudo-range because it is not a direct measurement of distance, but a measurement based on time. Pseudo-range is subject to several error sources; for example: ionospheric and tropospheric delays and multipath.

5. In addition to knowing the distance to a satellite, a receiver needs to know the satellite's exact position in space; this is known as its ephemeris. Each satellite transmits information about its exact orbital location. The GPS receiver uses this information to precisely establish the position of the satellite.

6. Using the calculated pseudo-range and position information supplied by the satellite, the GPS receiver mathematically determines its position by triangulation. The GPS receiver needs at least four satellites to yield a three-dimensional position (latitude, longitude, and altitude) and time solution. The GPS receiver computes navigational values such as distance and bearing to a waypoint, ground speed, etc., by using the aircraft's known latitude/longitude and referencing these to a database built into the receiver.

7. The GPS constellation of 24 satellites is designed so that a minimum of five are always observable by a user anywhere on Earth. The receiver uses data from a minimum of four satellites above the mask angle (the lowest angle above the horizon at which it can use a satellite).

8. The GPS receiver verifies the integrity (usability) of the signals received from the GPS constellation through receiver autonomous integrity monitoring (RAIM) to determine if a satellite is providing corrupted information. At least one satellite, in addition to those required for navigation, must be in view for

the receiver to perform the RAIM function; thus, RAIM needs a minimum of 5 satellites in view, or 4 satellites and a barometric altimeter (taro-aiding) to detect an integrity anomaly. For receivers capable of doing so, RAIM needs 6 satellites in view (or 5 satellites with baro-aiding) to isolate the corrupt satellite signal and remove it from the navigation solution. Baro-aiding is a method of augmenting the GPS integrity solution by using a nonsatellite input source. GPS derived altitude should not be relied upon to determine aircraft altitude since the vertical error can be quite large. To ensure that baro-aiding is available, the current altimeter setting must be entered into the receiver as described in the operating manual.

9. RAIM messages vary somewhat between receivers; however, generally there are two types. One type indicates that there are not enough satellites available to provide RAIM integrity monitoring and another type indicates that the RAIM integrity monitor has detected a potential error that exceeds the limit for the current phase of flight. **Without RAIM capability, the pilot has no assurance of the accuracy of the GPS position.**

10. The DOD declared initial operational capability (IOC) of the U.S. GPS on December 8, 1993. The FAA has granted approval for U.S. civil operators to use properly certified GPS equipment as a primary means of navigation in oceanic airspace and certain remote areas. Properly certified GPS equipment may be used as a supplemental means of IFR navigation for domestic en route, terminal operations, and certain instrument approach procedures (IAPs). This approval permits the use of GPS in a manner that is consistent with current navigation requirements as well as approved air carrier operations specifications.

b. VFR Use of GPS

1. GPS navigation has become a great asset to VFR pilots, providing increased navigation capability and enhanced situational awareness, while reducing operating costs due to greater ease in flying direct routes. While GPS has many benefits to the VFR pilot, care must be exercised to ensure that system capabilities are not exceeded.

2. Types of receivers used for GPS navigation under VFR are varied, from a full IFR installation being used to support a VFR flight, to a VFR only installation (in either a VFR or IFR capable aircraft) to a hand-held receiver. The limitations of each type of receiver installation or use must be understood by the pilot to avoid misusing navigation information. (See Table 1-1-8.) In all cases, VFR pilots should never rely solely on one system of navigation. GPS navigation must be integrated with other forms of electronic navigation (when possible), as well as pilotage and dead reckoning. Only through the integration of these techniques can the VFR pilot ensure accuracy in navigation.

3. Some critical concerns in VFR use of GPS include RAIM capability, database currency and antenna location.

(a) RAIM Capability. Many VFR GPS receivers and all hand-held units have no RAIM alerting capability. Loss of the required number of satellites in view, or the detection of a position error, cannot be displayed to the pilot by such receivers. In receivers with no RAIM capability, no alert would be provided to the pilot that the navigation solution had deteriorated, and an undetected navigation error could occur. A systematic cross-check with other navigation techniques would identify this failure, and prevent a serious deviation. See subparagraphs a8 and a9 for more information on RAIM.

(b) Database Currency

    (1) In many receivers, an up-datable database is used for navigation fixes, airports, and instrument procedures. These databases must be maintained to the current update for IFR operation, but no such requirement exists for VFR use.

    (2) However, in many cases, the database drives a moving map display which indicates Special Use Airspace and the various classes of airspace, in addition to other operational information. Without a current database the moving map display may be outdated and offer erroneous information to VFR pilots wishing to fly around critical airspace areas, such as a Restricted Area or a Class B airspace segment. Numerous pilots have ventured into airspace they were trying to avoid by using an outdated database. If you don't have a current database in the receiver, disregard the moving map display for critical navigation decisions.

    (3) In addition, waypoints are added, removed, relocated, or re-named as required to meet operational needs. When using GPS to navigate relative to a named fix, a current database must be used to properly locate a named waypoint. Without the update, it is the pilot's responsibility to verify the waypoint location referencing to an official current source, such as the Airport/Facility Directory, Sectional Chart, or En Route Chart.

(c) Antenna Location

    (1) In many VFR installations of GPS receivers, antenna location is more a matter of convenience than performance. In IFR installations, care is exercised to ensure that an adequate clear view is provided for the antenna to see satellites. If an alternate location is used, some portion of the aircraft may block the view of the antenna, causing a greater opportunity to lose navigation signal.

    (2) This is especially true in the case of handhelds. The use of hand-held receivers for VFR operations is a growing trend, especially among rental pilots. Typically, suction cups are used to place the GPS antennas on the inside of cockpit windows. While this method has great utility, the antenna location is limited to the cockpit or cabin only and is rarely optimized to provide a clear view of available satellites. Consequently, signal losses may occur in certain situations of aircraft-satellite

geometry, causing a loss of navigation signal. These losses, coupled with a lack of RAIM capability, could present erroneous position and navigation information with no warning to the pilot.

(3) While the use of a hand-held GPS for VFR operations is not limited by regulation, modification of the aircraft, such as installing a panel- or yoke-mounted holder, is governed by 14 CFR Part 43. Consult with your mechanic to ensure compliance with the regulation, and a safe installation.

4. As a result of these and other concerns, here are some tips for using GPS for VFR operations:

(a) Always check to see if your unit has RAIM capability. If no RAIM capability exists, be suspicious of your GPS position when any disagreement exists with the position derived from other radio navigation systems, pilotage, or dead reckoning.

(b) Check the currency of the database, if any. If expired, update the database using the current revision. If an update of an expired database is not possible, disregard any moving map display of airspace for critical navigation decisions. Be aware that named waypoints may no longer exist or may have been relocated since the database expired. At a minimum, the waypoints planned to be used should be checked against a current official source, such as the Airport/Facility Directory, or a Sectional Aeronautical Chart.

(c) While hand-helds can provide excellent navigation capability to VFR pilots, be prepared for intermittent loss of navigation signal, possibly with no RAIM warning to the pilot. If mounting the receiver in the aircraft, be sure to comply with 14 CFR Part 43.

(d) Plan flights carefully before taking off. If you wish to navigate to user-defined waypoints, enter them before flight, not on-the-fly. Verify your planned flight against a current source, such as a current sectional chart. There have been cases in which one pilot used waypoints created by another pilot that were not where the pilot flying was expecting. This generally resulted in a navigation error. Minimize head-down time in the aircraft and keep a sharp lookout for traffic, terrain, and obstacles. Just a few minutes of preparation and planning on the ground will make a great difference in the air.

(e) Another way to minimize head-down time is to become very familiar with your receiver's operation. Most receivers are not intuitive. The pilot must take the time to learn the various keystrokes, knob functions, and displays that are used in the operation of the receiver. Some manufacturers provide computer-based tutorials or simulations of their receivers. Take the time to learn about your particular unit before you try to use it in flight.

5. In summary, be careful not to rely on GPS to solve all your VFR navigational problems. Unless an IFR receiver is installed in accordance with IFR requirements, no standard of accuracy or integrity has been assured. While the practicality of GPS is compelling, the fact remains that only the pilot can navigate the aircraft, and GPS is just one of the pilot's tools to do the job.

c. VFR Waypoints

1. VFR waypoints provide VFR pilots with a supplementary tool to assist with position awareness while navigating visually in aircraft equipped with area navigation receivers. VFR waypoints should be used as a tool to supplement current navigation procedures. The uses of VFR waypoints include providing navigational aids for pilots unfamiliar with an area, waypoint definition of existing reporting points, enhanced navigation in and around Class B and Class C airspace, and enhanced navigation around Special Use Airspace. VFR pilots should rely on appropriate and current aeronautical charts published specifically for visual navigation. If operating in a terminal area, pilots should take advantage of the Terminal Area Chart available for that area, if published. The use of VFR waypoints does not relieve the pilot of any responsibility to comply with the operational requirements of 14 CFR Part 91.

2. VFR waypoint names (for computer-entry and flight plans) consist of five letters beginning with the letters "VP" and are retrievable from navigation databases. **NOTICE: Effective on 6/15/00 VFR waypoint names shall consist of five letters beginning with the letters "VP." The change is effective for all GPS databases and aviation publications. The Los Angeles Helicopter Route Chart depicts VFR waypoint names beginning with "W." The chart will be updated to the "VP" naming convention at the next publication of the chart.** The VFR waypoint names are not intended to be pronounceable, and they are not for use in ATC communications. On VFR charts, stand-alone VFR waypoints will be portrayed using the same four-point star symbol used for IFR waypoints. VFR waypoints collocated with visual check points on the chart will be identified by small magenta flag symbols. VFR waypoints collocated with visual check points will be pronounceable based on the name of the visual check point and may be used for ATC communications. Each VFR waypoint name will appear in parentheses adjacent to the geographic location on the chart. Latitude/longitude data for all established VFR waypoints may be found in the appropriate regional Airport/Facility Directory (A/FD).

3. VFR waypoints shall not be used to plan flights under IFR. VFR waypoints will not be recognized by the IFR system and will be rejected for IFR routing purposes.

4. When filing VFR flight plans, pilots may use the five letter identifier as a waypoint in the route of flight section if there is an intended course change at that point or if used to describe the planned route of flight. This VFR filing would be similar to how a VOR would be used in a route of flight. Pilots must use the VFR waypoints only when operating under VFR conditions.

5. Any VFR waypoints intended for use during a flight should be loaded into the receiver while on the ground and prior to departure. Once airborne, pilots should avoid programming routes or VFR waypoint chains into their receivers.

6. Pilots should be especially vigilant for other traffic while operating near VFR

waypoints. The same effort to see and avoid other aircraft near VFR waypoints will be necessary, as was the case with VORs and NDBs in the past. In fact, the increased accuracy of navigation through the use of GPS will demand even greater vigilance, as off-course deviations among different pilots and receivers will be less. When operating near a VFR waypoint, use whatever ATC services are available, even if outside a class of airspace where communications are required. Regardless of the class of airspace, monitor the available ATC frequency closely for information on other aircraft operating in the vicinity. It is also a good idea to turn on your landing light(s) when operating near a VFR waypoint to make your aircraft more conspicuous to other pilots, especially when visibility is reduced. See paragraph 7-5-2, VFR in Congested Areas, for more information.

d. The Gulf of Mexico Grid System

1. On October 8, 1998, the Southwest Region of the FAA, with assistance from the Helicopter Safety Advisory Conference (HSAC), implemented the world's first Instrument Flight Rules (IFR) Grid System in the Gulf of Mexico. This navigational route structure is completely independent of ground-based navigation aids (NAVAIDs) and was designed to facilitate helicopter IFR operations to offshore destinations. The Grid System is defined by over 300 offshore waypoints located 20 minutes apart (latitude and longitude). Flight plan routes are routinely defined by just 4 segments; departure point (lat/long), first en route grid waypoint, last en route grid waypoint prior to approach procedure, and destination point (lat/long). There are over 4,000 possible offshore landing sites. Upon reaching the waypoint prior to the destination, the pilot may execute an Offshore Standard Approach Procedure (OSAP), a Helicopter En Route Descent Areas (HEDA) approach, or an Airborne Radar Approach (ARA). For more information on these helicopter instrument procedures, refer to FAA AC 90-80B, Approval of Offshore Standard Approach Procedure (OSAP), Airborne Radar Approaches (ARA), and Helicopter En Route Areas (HEDA) Criteria, on the Flight Standards web site http://terps.faa.gov. The return flight plan is just the reverse with the requested stand-alone GPS approach contained in the remarks section.

2. The large number (over 300) of waypoints in the grid system makes it difficult to assign phonetically pronounceable names to the waypoints that would be meaningful to pilots and controllers. A unique naming system was adopted that enables pilots and controllers to derive the fix position from the name. The five-letter names are derived as follows:

   (a) The waypoints are divided into sets of 3 columns each. A three-letter identifier, identifying a geographical area or a NAVAID to the north, represents each set.

   (b) Each column in a set is named after its position, i.e., left (L), center (C), and right (R).

   (c) The rows of the grid are named alphabetically from north to south, starting with A for the northern most row.

Example: *LCHRC would be pronounced "Lake Charles Romeo Charlie." The way-point is in the right-hand column of the Lake Charles VOR set, in row C (third south from the northern most row).*

3. Since the grid system's implementation, IFR delays (frequently over 1 hour in length) for operations in this environment have been effectively eliminated. The comfort level of the pilots, knowing that they will be given a clearance quickly, plus the mileage savings in this near free-flight environment, is allowing the operators to carry less fuel. Less fuel means they can transport additional passengers, which is a substantial fiscal and operational benefit, considering the limited seating on board helicopters.

4. There are 3 requirements for operators to meet before filing IFR flight plans utilizing the grid:
   (a) The helicopter must be IFR certified and equipped with IFR certified TSO C-129 GPS navigational units.
   (b) The operator must obtain prior written approval from the appropriate Flight Standards District Office through a Certificate of Authorization or revision to their Operations Specifications, as appropriate.
   (c) The operator must be a signatory to the Houston ARTCC Letter of Agreement.

5. FAA/NACO publishes the grid system waypoints on the IFR Gulf of Mexico Vertical Flight Reference Chart. A commercial equivalent is also available. The chart is updated annually and is available from a FAA chart agent or FAA directly, website address: http://acc.nos.noaa.gov.

e. General Requirements

1. Authorization to conduct any GPS operation under IFR requires that:
   (a) GPS navigation equipment used must be approved in accordance with the requirements specified in Technical Standard Order (TSO) C-129, or equivalent, and the installation must be done in accordance with Advisory Circular AC 20-138, Airworthiness Approval of Global Positioning System (GPS) Navigation Equipment for Use as a VFR and IFR Supplemental Navigation System, or Advisory Circular AC 20-130A, Airworthiness Approval of Navigation or Flight Management Systems Integrating Multiple Navigation Sensors, or equivalent. Equipment approved in accordance with TSO C-1 15a does not meet the requirements of TSO C-129. Visual flight rules (VFR) and hand-held GPS systems are not authorized for IFR navigation, instrument approaches, or as a principal instrument flight reference. During IFR operations they may be considered only an aid to situational awareness.
   (b) Aircraft using GPS navigation equipment under IFR must be equipped with an approved and operational alternate means of navigation appropriate to the flight. Active monitoring of alternative navigation equipment is not required if the GPS receiver uses RAIM for integrity monitoring.

Active monitoring of an alternate means of navigation is required when the RAIM capability of the GPS equipment is lost.

(c) Procedures must be established for use in the event that the loss of RAIM capability is predicted to occur. In situations where this is encountered, the flight must rely on other approved equipment, delay departure, or cancel the flight.

(d) The GPS operation must be conducted in accordance with the FAA-approved aircraft flight manual (AFM) or flight manual supplement. Flight crew members must be thoroughly familiar with the particular GPS equipment installed in the aircraft, the receiver operation manual, and the AFM or flight manual supplement. Unlike ILS and VOR, the basic operation, receiver presentation to the pilot, and some capabilities of the equipment can vary greatly. Due to these differences, operation of different brands, or even models of the same brand, of GPS receiver under IFR should not be attempted without thorough study of the operation of that particular receiver and installation. Most receivers have a built-in simulator mode which will allow the pilot to become familiar with operation prior to attempting operation in the aircraft. Using the equipment in flight under VFR conditions prior to attempting 11R operation will allow further familiarization.

(e) Aircraft navigating by IFR approved GPS are considered to be area navigation (RNAV) aircraft and have special equipment suffixes. File the appropriate equipment suffix in accordance with TBL 5-1-2, on the ATC flight plan. If GPS avionics become inoperative, the pilot should advise ATC and amend the equipment suffix.

(f) Prior to any GPS IFR operation, the pilot must review appropriate NOTAMs and aeronautical information. (See GPS NOTAMs/Aeronautical Information.)

(g) Air carrier and commercial operators must meet the appropriate provisions of their approved operations specifications.

f. Use of GPS for IFR Oceanic, Domestic En Route, and Terminal Area Operations

1. GPS IFR operations in oceanic areas can be conducted as soon as the proper avionics systems are installed, provided all general requirements are met. A GPS installation with TSO C-129 authorization in class A1, A2, B1, B2, C1, or C2 may be used to replace one of the other approved means of long-range navigation, such as dual INS or dual Omega. (See TBL 1-1-7 and TBL 1-1-8.) A single GPS installation with these classes of equipment which provide RAIM for integrity monitoring may also be used on short oceanic routes which have only required one means of long-range navigation.

2. GPS domestic en route and terminal IFR operations can be conducted as soon as proper avionics systems are installed, provided all general requirements are met. The avionics necessary to receive all of the ground-based facilities appro-

TBL 1-1-7
GPS IFR Equipment Classes/Categories

Equipment Class	RAIM	Int. Nav Sys. to Prov. RAIM Equiv.	Oceanic	En Route	Terminal	Nonprecision Approach Capable
TSO-C129						
Class A - GPS sensor and navigation capability.						
A1	yes		yes	yes	yes	yes
A2	yes		yes	yes	yes	no
Class B - GPS sensor data to an integrated navigation system (i.e. FMS, multi-sensor navigation system, etc.).						
B1	yes		yes	yes	yes	yes
B2	yes		yes	yes	yes	no
B3		yes	yes	yes	yes	yes
B4		yes	yes	yes	yes	no
Class C - GPS sensor data to an integrated navigation system (as in Class B) which provides enhanced guidance to an autopilot, or flight director, to reduce flight tech. errors. Limited to 14 CFR Part 121 or equivalent criteria.						
C1	yes		yes	yes	yes	yes
C2	yes		yes	yes	yes	no
C3		yes	yes	yes	yes	yes
C4		yes	yes	yes	yes	no

TBL 1-1-8
GPS Approval Required/Authorized Use

Equipment Type[1]	Installation Approval Required	Operational Approval Required	IFR En Route[2]	IFR Terminal[2]	IFR Approach[3]	Oceanic Remote	In Lieu of ADF and/or DME[3]
Hand held[4]	X[5]						
VFR Panel Mount[4]	X						
IFR En Route and Terminal	X	X	X	X			X
IFR Oceanic/ Remote	X	X	X	X		X	X
IFR En Route, Terminal, and Approach	X	X	X	X	X		X

NOTE-
[1]*To determine equipment approvals and limitations, refer to the AFM, AFM supplements, or pilot guides.*
[2]*Requires verification of data for correctness if database is expired.*
[3]*Requires current database.*
[4]*VFR and hand-held GPS systems are not authorized for IFR navigation, instrument approaches, or as a primary instrument flight reference. During IFR operations they may be considered only an aid to situational awareness.*
[5]*Hand-held receivers require no approval. However, any aircraft modification to support the hand-held receiver; i.e., installation of an external antenna or a permanent mounting bracket, does require approval.*

priate for the route to the destination airport and any required alternate airport must be installed and operational. Ground-based facilities necessary for these routes must also be operational.

(a) GPS en route IFR RNAV operations may be conducted in Alaska outside the operational service volume of ground-based navigation aids when a TSO-C145 or TSO-146a GPS/WAAS system is installed and operating. Ground-based navigation equipment is not required to be installed and operating for en route IFR/RNAV operations when using GPS WAAS navigation systems. All operators should ensure that an alternate means of navigation is available in the unlikely event the GPS WAAS navigation system becomes inoperative.

3. The GPS Approach Overlay Program is an authorization for pilots to use GPS avionics under IFR for flying designated nonprecision instrument approach procedures, except LOC, LDA, and simplified directional facility (SDF) procedures. These procedures are now identified by the name of the procedure and "or GPS" (e.g., VOR/DME or GPS RWY 15). Other previous types of overlays have either been converted to this format or replaced with stand-alone procedures. Only approaches contained in the current onboard navigation database are authorized. The navigation database may contain information about nonoverlay approach procedures that is intended to be used to enhance position orientation, generally by providing a map, while flying these approaches using conventional NAVAIDs. This approach information should not be confused with a GPS overlay approach (see the receiver operating manual, AFM, or AFM Supplement for details on how to identify these approaches in the navigation database).

Note: *Overlay approaches are predicated upon the design criteria of the ground-based NAVAID used as the basis of the approach. As such, they do not adhere to the design criteria described in paragraph 5-4-5; Area Navigation (RNAV) Instrument Approach Charts, for stand-alone GPS approaches.*

4. GPS IFR approach operations can be conducted as soon as proper avionics systems are installed and the following requirements are met:
   (a) The authorization to use GPS to fly instrument approaches is limited to U.S. airspace.
   (b) The use of GPS in any other airspace must be expressly authorized by the FAA Administrator.
   (c) GPS instrument approach operations outside the U.S. must be authorized by the appropriate sovereign authority.

5. Subject to the restrictions below, operators in the U.S. NAS are authorized to use GPS equipment certified for IFR operations in place of ADF and/or DME equipment for en route and terminal operations. For some operations there is no requirement for the aircraft to be equipped with an ADF or DME receiver, see subparagraphs 6(g) and (h) below. The ground-based NDB or DME facility may be temporarily out of service during these operations. Charting will not change to support these operations.
   (a) Determining the aircraft position over a DME fix. GPS satisfies the 14 CFR Section 91.205(e) requirement for DME at and above 24,000 feet mean sea level (MSL) (FL 240).
   (b) Flying a DME arc.
   (c) Navigating to/from an NDB/compass locator.
   (d) Determining the aircraft position over an NDB/compass locator.
   (e) Determining the aircraft position over a fix defined by an NDB/compass locator bearing crossing a VOR/LOC course.

(f) Holding over an NDB/compass locator.

Note: *This approval does not alter the conditions and requirements for use of GPS to fly existing nonprecision instrument approach procedures as defined in the GPS approach overlay program.*

6. Restrictions

   (a) GPS avionics approved for terminal IFR operations may be used in lieu of ADF and/or DME. Included in this approval are both stand-alone and multi-sensor systems actively employing GPS as a sensor. This equipment must be installed in accordance with appropriate airworthiness installation requirements and the provisions of the applicable FAA approved AFM, AFM supplement, or pilot's guide must be met. The required integrity for these operations must be provided by at least en route RAIM, or an equivalent method; i.e., Wide Area Augmentation System (WAAS).

   (b) For air carriers and operators for compensation or hire, Principal Operations Inspector (POI) and operations specification approval is required for any use of GPS.

   (c) Waypoints, fixes, intersections, and facility locations to be used for these operations must be retrieved from the GPS airborne database. The database must be current. If the required positions cannot be retrieved from the airborne database, the substitution of GPS for ADF and/or DME is not authorized.

   (d) The aircraft GPS system must be operated within the guidelines contained in the AFM, AFM supplement, or pilot's guide.

   (e) The CDI must be set to terminal sensitivity (normally 1 or 1 1/4 NM) when tracking GPS course guidance in the terminal area. This is to ensure that small deviations from course are displayed to the pilot in order to keep the aircraft within the smaller terminal protected areas.

   (f) Charted requirements for ADF and/or DME can be met using the GPS system, except for use as the principal instrument approach navigation source.

   (g) Procedures must be established for use in the event that GPS integrity outages are predicted or occur (RAIM annunciation). In these situations, the flight must rely on other approved equipment; this may require the aircraft to be equipped with operational NDB and/or DME receivers. Otherwise, the flight must be rerouted, delayed, canceled or conducted VFR.

   (h) A non-GPS approach procedure must exist at the alternate airport when one is required. If the non-GPS approaches on which the pilot must rely require DME or ADF, the aircraft must be equipped with DME or ADF avionics as appropriate.

7. Guidance. The following provides general guidance which is not specific to any particular aircraft GPS system. For specific system guidance refer to the AFM, AFM supplement, pilot's guide, or contact the manufacturer of your system.

   (a) To determine the aircraft position over a DME fix:

(1) Verify aircraft GPS system integrity monitoring is functioning properly and indicates satisfactory integrity.

(2) If the fix is identified by a five letter name which is contained in the GPS airborne database, you may select either the named fix as the active GPS waypoint (WP) or the facility establishing the DME fix as the active GPS WP.

Note: *When using a facility as the active WP, the only acceptable facility is the DME facility which is charted as the one used to establish the DME fix. If this facility is not in your airborne database, you are not authorized to use a facility WP for this operation.*

(3) If the fix is identified by a five letter name which is not contained in the GPS airborne database, or if the fix is not named, you must select the facility establishing the DME fix or another named DME fix as the active GPS WP.

Note: *An alternative, until all DME sources are in the database, is using a named DME fix as the active waypoint to identify unnamed DME fixes on the same course and from the same DME source as the active waypoint.*

Caution: *Pilots should be extremely careful to ensure that correct distance measurements are used when utilizing this interim method. It is strongly recommended that pilots review distances for DME fixing during preflight preparation.*

(4) If you select the named fix as your active GPS WP, you are over the fix when the GPS system indicates you are at the active WP.

(5) If you select the DME providing facility as the active GPS WP, you are over the fix when the GPS distance from the active WP equals the charted DME value and you are on the appropriate bearing or course.

(b) To fly a DME arc:

(1) Verify aircraft GPS system integrity monitoring is functioning properly and indicates satisfactory integrity.

(2) You must select, from the airborne database, the facility providing the DME arc as the active GPS WP.

Note: *The only acceptable facility is the DME facility on which the arc is based. If this facility is not in your airborne database, you are not authorized to perform this operation.*

(3) Maintain position on the arc by reference to the GPS distance in lieu of a DME readout.

(c) To navigate to or from an NDB/compass locator:

Note: *If the chart depicts the compass locator collocated with a fix of the same name, use of that fix as the active WP in place of the compass locator facility is authorized.*

(1) Verify aircraft GPS system integrity monitoring is functioning properly and indicates satisfactory integrity.

(2) Select terminal CDI sensitivity in accordance with the AFM, AFM supplement, or pilot's guide if in the terminal area.

(3) Select the NDB/compass locator facility from the airborne database as the active WP.

(4) Select and navigate on the appropriate course to or from the active WP.

(d) To determine the aircraft position over an NDB/compass locator:

(1) Verify aircraft GPS system integrity monitoring is functioning properly and indicates satisfactory integrity.

(2) Select the NDB/compass locator facility from the airborne database as the active WP.

Note: *When using an NDB/compass locator, that facility must be charted and be in the airborne database. If this facility is not in your airborne database, you are not authorized to use a facility WP for this operation.*

(3) You are over the NDB/compass locator when the GPS system indicates you are at the active WP.

(e) To determine the aircraft position over a fix made up of an NDB/compass locator bearing crossing a VOR/LOC course:

(1) Verify aircraft GPS system integrity monitoring is functioning properly and indicates satisfactory integrity.

(2) A fix made up by a crossing NDB/compass locator bearing will be identified by a five letter fix name. You may select either the named fix or the NDB/compass locator facility providing the crossing bearing to establish the fix as the active GPS WP.

Note: *When using an NDB/compass locator, that facility must be charted and be in the airborne database. If this facility is not in your airborne database, you are not authorized to use a facility WP for this operation.*

(3) If you select the named fix as your active GPS WP, you are over the fix when the GPS system indicates you are at the WP as you fly the prescribed track from the non–GPS navigation source.

(4) If you select the NDB/compass locator facility as the active GPS WP, you are over the fix when the GPS bearing to the active WP is the same as the charted NDB/compass locator bearing for the fix as you fly the prescribed track from the non–GPS navigation source.

(f) To hold over an NDB/compass locator:

(1) Verify aircraft GPS system integrity monitoring is functioning properly and indicates satisfactory integrity.

(2) Select terminal CDI sensitivity in accordance with the AFM, AFM supplement, or pilot's guide if in the terminal area.

(3) Select the NDB/compass locator facility from the airborne database as the active WP.

Note: *When using a facility as the active WP, the only acceptable facility is the NDB/compass locator facility which is charted. If this facility is not in your airborne database, you are not authorized to use a facility WP for this operation.*

(4) Select nonsequencing (e.g. "HOLD" or "OBS") mode and the appropriate course in accordance with the AFM, AFM supplement, or pilot's guide.

(5) Hold using the GPS system in accordance with the AFM, AFM supplement, or pilot's guide.

8. Planning. Good advance planning and intimate knowledge of your navigational systems are vital to safe and successful use of GPS in lieu of ADF and/or DME.

(a) You should plan ahead before using GPS systems as a substitute for ADF and/or DME. You will have several alternatives in selecting waypoints and system configuration. After you are cleared for the approach is not the time to begin programming your GPS. In the flight planning process you should determine whether you will use the equipment in the automatic sequencing mode or in the nonsequencing mode and select the waypoints you will use.

(b) When you are using your aircraft GPS system to supplement other navigation systems, you may need to bring your GPS control panel into your navigation scan to see the GPS information. Some GPS aircraft installations will present localizer information on the CDI whenever a localizer frequency is tuned, removing the GPS information from the CDI display. Good advance planning and intimate knowledge of your navigation systems are vital to safe and successful use of GPS.

(c) The following are some factors to consider when preparing to install a GPS receiver in an aircraft. Installation of the equipment can determine how easy or how difficult it will be to use the system.

(1) Consideration should be given to installing the receiver within the primary instrument scan to facilitate using the GPS in lieu of ADF and/or DME. This will preclude breaking the primary instrument scan while flying the aircraft and tuning, and identifying waypoints. This becomes increasingly important on approaches, and missed approaches.

(2) Many GPS receivers can drive an ADF type bearing pointer. Such an installation will provide the pilot with an enhanced level of situational awareness by providing GPS navigation information while the CDI is set to VOR or ILS.

(3) The GPS receiver may be installed so that when an ILS frequency is tuned, the navigation display defaults to the VOR/ILS mode, preempting the GPS mode. However, if the receiver installation requires a manual selection from GPS to ILS, it allows the ILS to be tuned and identified while navigating on the GPS. Additionally, this prevents the navigation display from automatically switching back to GPS when a VOR frequency is selected. If the navigation display automatically

switches to GPS mode when a VOR is selected, the change may go unnoticed and could result in erroneous navigation and departing obstruction protected airspace.

(4) GPS is a supplemental navigation system in part due to signal availability. There will be times when your system will not receive enough satellites with proper geometry to provide accurate positioning or sufficient integrity. Procedures should be established by the pilot in the event that GPS outages occur. In these situations, the pilot should rely on other approved equipment, delay departure, reroute, or discontinue IFR operations.

g. Equipment and Database Requirements

1. Authorization to fly approaches under IFR using GPS avionics systems requires that:

   (a) A pilot use GPS avioniczs with TSO C-129, or equivalent, authorization in class A1, B1, B3, C1, or C3; and

   (b) All approach procedures to be flown must be retrievable from the current airborne navigation database supplied by the TSO C-129 equipment manufacturer or other FAA approved source.

h. GPS Approach Procedures

As the production of stand-alone GPS approaches has progressed, many of the original overlay approaches have been replaced with stand-alone procedures specifically designed for use by GPS systems. The title of the remaining GPS overlay procedures has been revised on the approach chart to "or GPS" (e.g., VOR or GPS RWY 24). Therefore, all the approaches that can be used by GPS now contain "GPS" in the title (e.g., "VOR or GPS RWY 24," "GPS RWY 24," or "RNAV (GPS) RWY 24"). During these GPS approaches, underlying ground-based NAVAIDs are not required to be operational and associated aircraft avionics need not be installed, operational, turned on or monitored (monitoring of the underlying approach is suggested when equipment is available and functional). Existing overlay approaches may be requested using the GPS title, such as "GPS RWY 24" for the VOR or GPS RWY 24.

Note: *Any required alternate airport must have an approved instrument approach procedure other than GPS that is anticipated to be operational and available at the estimated time of arrival, and which the aircraft is equipped to fly.*

i. GPS NOTAMs/Aeronautical Information

1. GPS satellite outages are issued as GPS NOTAMs both domestically and internationally. However, the effect of an outage on the intended operation cannot be determined unless the pilot has a RAIM availability prediction program which allows excluding a satellite which is predicted to be out of service based on the NOTAM information.

2. Civilian pilots may obtain GPS RAIM availability information for nonprecision

approach procedures by specifically requesting GPS aeronautical information from an Automated Flight Service Station during preflight briefings. GPS RAIM aeronautical information can be obtained for a period of 3 hours (ETA hour and 1 hour before to 1 hour after the ETA hour) or a 24 hour time frame at a particular airport. FAA briefers will provide RAIM information for a period of 1 hour before to 1 hour after the ETA, unless a specific time frame is requested by the pilot. If flying a published GPS departure, a RAIM prediction should also be requested for the departure airport.

3. The military provides airfield specific GPS RAIM NOTAMs for nonprecision approach procedures at military airfields. The RAIM outages are issued as M-series NOTAMs and may be obtained for up to 24 hours from the time of request.

4. Receiver manufacturers and/or database suppliers may supply "NOTAM" type information concerning database errors. Pilots should check these sources, when available, to ensure that they have the most current information concerning their electronic database.

j. Receiver Autonomous Integrity Monitoring (RAIM)

1. RAIM outages may occur due to an insufficient number of satellites or due to unsuitable satellite geometry which causes the error in the position solution to become too large. Loss of satellite reception and RAIM warnings may occur due to aircraft dynamics (changes in pitch or bank angle). Antenna location on the aircraft, satellite position relative to the horizon, and aircraft attitude may affect reception of one or more satellites. Since the relative positions of the satellites are constantly changing, prior experience with the airport does not guarantee reception at all times, and RAIM availability should always be checked.

2. If RAIM is not available, another type of navigation and approach system must be used, another destination selected, or the trip delayed until RAIM is predicted to be available on arrival. On longer flights, pilots should consider rechecking the RAIM prediction for the destination during the flight. This may provide early indications that an unscheduled satellite outage has occurred since takeoff.

3. If a RAIM failure/status annunciation occurs prior to the final approach way-point (FAWP), the approach should not be completed since GPS may no longer provide the required accuracy. The receiver performs a RAIM prediction by 2 NM prior to the FAWP to ensure that RAIM is available at the FAWP as a condition for entering the approach mode. **The pilot should ensure that the receiver has sequenced from "Armed" to "Approach" prior to the FAWP** (normally occurs 2 NM prior). Failure to sequence may be an indication of the detection of a satellite anomaly, failure to arm the receiver (if required), or other problems which preclude completing the approach.

4. If the receiver does not sequence into the approach mode or a RAIM failure/status annunciation occurs prior to the FAWP, the pilot should not

descend to Minimum Descent Altitude (MDA), but should proceed to the missed approach waypoint (MAWP) via the FAWP, perform a missed approach, and contact ATC as soon as practical. Refer to the receiver operating manual for specific indications and instructions associated with loss of RAIM prior to the FAF.

5. If a RAIM failure occurs after the FAWP, the receiver is allowed to continue operating without an annunciation for up to 5 minutes to allow completion of the approach (see receiver operating manual). **If the RAIM flag/status annunciation appears after the FAWP, the missed approach should be executed immediately.**

k. Waypoints

1. GPS approaches make use of both fly-over and fly-by waypoints. Fly-by waypoints are used when an aircraft should begin a turn to the next course prior to reaching the waypoint separating the two route segments. This is known as turn anticipation and is compensated for in the airspace and terrain clearances. Approach waypoints, except for the MAWP and the missed approach holding waypoint (MAHWP), are normally fly-by waypoints. Fly-over waypoints are used when the aircraft must fly over the point prior to starting a turn. New approach charts depict fly-over waypoints as a circled waypoint symbol. Overlay approach charts and some early stand alone GPS approach charts may not reflect this convention.

2. Since GPS receivers are basically "To–To" navigators, they must always be navigating to a defined point. On overlay approaches, if no pronounceable five-character name is published for an approach waypoint or fix, it was given a database identifier consisting of letters and numbers. These points will appear in the list of waypoints in the approach procedure database, but may not appear on the approach chart. A point used for the purpose of defining the navigation track for an airborne computer system (i.e., GPS or FMS) is called a Computer Navigation Fix (CNF). CNFs include unnamed DME fixes, beginning and ending points of DME arcs and sensor final approach fixes (FAFs) on some GPS overlay approaches. To aid in the approach chart/database correlation process, the FAA has begun a program to assign five-letter names to CNFs and to chart CNFs on various National Oceanic Service aeronautical products. These CNFs are not to be used for any air traffic control (ATC) application, such as holding for which the fix has not already been assessed. CNFs will be charted to distinguish them from conventional reporting points, fixes, intersections, and waypoints. The CNF name will be enclosed in parenthesis, e.g., (MABEE), and the name will be placed next to the CNF it defines. If the CNF is not at an existing point defined by means such as crossing radials or radial/DME, the point will be indicated by an "X." The CNF name will not be used in filing a flight plan or in aircraft/ATC communications. Use current phraseology, e.g., facility name, radial, distance, to describe these fixes.

3. Unnamed waypoints in the database will be uniquely identified for each airport but may be repeated for another airport (e.g., RW36 will be used at each airport with a runway 36 but will be at the same location for all approaches at a given airport).
4. The runway threshold waypoint, which is normally the MAWP, may have a five letter identifier (e.g., SNEEZ) or be coded as RW## (e.g., RW36, RW36L). Those thresholds which are coded as five letter identifiers are being changed to the RW## designation. This may cause the approach chart and database to differ until all changes are complete. The runway threshold waypoint is also used as the center of the Minimum Safe Altitude (MSA) on most GPS approaches. MAWPs not located at the threshold will have a five letter identifier.

l. Position Orientation: As with most RNAV systems, pilots should pay particular attention to position orientation while using GPS. Distance and track information are provided to the next active waypoint, not to a fixed navigation aid. Receivers may sequence when the pilot is not flying along an active route, such as when being vectored or deviating for weather, due to the proximity to another waypoint in the route. This can be prevented by placing the receiver in the nonsequencing mode. When the receiver is in the nonsequencing mode, bearing and distance are provided to the selected waypoint and the receiver will not sequence to the next waypoint in the route until placed back in the auto sequence mode or the pilot selects a different waypoint. On overlay approaches, the pilot may have to compute the along track distance to stepdown fixes and other points due to the receiver showing along track distance to the next waypoint rather than DME to the VOR or ILS ground station.

m. Conventional Versus GPS Navigation Data: There may be slight differences between the heading information portrayed on navigational charts and the GPS navigation display when flying an overlay approach or along an airway. All magnetic tracks defined by a VOR radial are determined by the application of magnetic variation at the VOR; however, GPS operations may use an algorithm to apply the magnetic variation at the current position, which may produce small differences in the displayed course. Both operations should produce the same desired ground track. Due to the use of great circle courses, and the variations in magnetic variation, the bearing to the next waypoint and the course from the last waypoint (if available) may not be exactly 180° apart when long distances are involved. Variations in distances will occur since GPS distance-to-waypoint values are along track (straight-line) distances (ATD) computed to the next waypoint and the DME values published on underlying procedures are slant range distances measured to the station. This difference increases with aircraft altitude and proximity to the NAVAID.

n. Departures and Instrument Departure Procedures (DPs): The GPS receiver must be set to terminal (±1 NM) CDI sensitivity and the navigation routes contained in the database in order to fly published IFR charted departures and DPs. Terminal

RAIM should be automatically provided by the receiver. (Terminal RAIM for departure may not be available unless the waypoints are part of the active flight plan rather than proceeding direct to the first destination.) Certain segments of a DP may require some manual intervention by the pilot, especially when radar vectored to a course or required to intercept a specific course to a waypoint. The database may not contain all of the transitions or departures from all runways and some GPS receivers do not contain DPs in the database. It is necessary that helicopter procedures be flown at 70 knots or less since helicopter departure procedures and missed approaches use a 20:1 obstacle clearance surface (OCS), which is double the fixed-wing OCS, and turning areas are based on this speed as well.

o. Flying GPS Approaches

1. Determining which area of the TAA the aircraft will enter when flying a "T" with a TAA must be accomplished using the bearing and distance to the IF(IAF). This is most critical when entering the TAA in the vicinity of the extended runway centerline and determining whether you will be entering the right or left base area. Once inside the TAA, all sectors and stepdowns are based on the bearing and distance to the IAF for that area, which the aircraft should be proceeding direct to at that time, unless on vectors. (See Fig. 5-4-3 and Fig. 5-4-4.)

2. Pilots should fly the full approach from an Initial Approach Waypoint (IAWP) or feeder fix unless specifically cleared otherwise. Randomly joining an approach at an intermediate fix does not assure terrain clearance.

3. When an approach has been loaded in the flight plan, GPS receivers will give an "arm" annunciation 30 NM straight line distance from the airport/heliport reference point. Pilots should arm the approach mode at this time, if it has not already been armed (some receivers arm automatically). Without arming, the receiver will not change from en route CDI and RAIM sensitivity of ±5 NM either side of centerline to ±1 NM terminal sensitivity. Where the IAWP is inside this 30 mile point, a CDI sensitivity change will occur once the approach mode is armed and the aircraft is inside 30 NM. Where the IAWP is beyond 30 NM from the airport/heliport reference point, CDI sensitivity will not change until the aircraft is within 30 miles of the airport/heliport reference point even if the approach is armed earlier. Feeder route obstacle clearance is predicated on the receiver being in terminal (±1 NM) CDI sensitivity and RAIM within 30 NM of the airport/heliport reference point, therefore, the receiver should always be armed (if required) not later than the 30 NM annunciation.

4. The pilot must be aware of what bank angle/turn rate the particular receiver uses to compute turn anticipation, and whether wind and airspeed are included in the receiver's calculations. This information should be in the receiver operating manual. Over or under banking the turn onto the final approach course may significantly delay getting on **course and may result in high descent rates to achieve the next segment altitude.**

5. When within 2 NM of the FAWP with the approach mode armed, the approach mode will switch to active, which results in RAIM changing to approach sensitivity and a change in CDI sensitivity. Beginning 2 NM prior to the FAWP, the full scale CDI sensitivity will smoothly change from ±1 NM to ±0.3 NM at the FAWP. As sensitivity changes from ±1 NM to ±0.3 NM approaching the FAWP, with the CDI not centered, the corresponding increase in CDI displacement may give the impression that the aircraft is moving further away from the intended course even though it is on an acceptable intercept heading. Referencing the digital track displacement information (cross track error), if it is available in the approach mode, may help the pilot remain position oriented in this situation. Being established on the final approach course prior to the beginning of the sensitivity change at 2 NM will help prevent problems in interpreting the CDI display during ramp down. Therefore, requesting or accepting vectors which will cause the aircraft to intercept the final approach course within 2 NM of the FAWP is not recommended.

6. When receiving vectors to final, most receiver operating manuals suggest placing the receiver in the nonsequencing mode on the FAWP and manually setting the course. This provides an extended final approach course in cases where the aircraft is vectored onto the final approach course outside of any existing segment which is aligned with the runway. Assigned altitudes must be maintained until established on a published segment of the approach. Required altitudes at waypoints outside the FAWP or stepdown fixes must be considered. Calculating the distance to the FAWP may be required in order to descend at the proper location.

7. Overriding an automatically selected sensitivity during an approach will cancel the approach mode annunciation. If the approach mode is not armed by 2 NM prior to the FAWP, the approach mode will not become active at 2 NM prior to the FAWP, and the equipment will flag. In these conditions, the RAIM and CDI sensitivity will not ramp down, and the pilot should not descend to MDA, but fly to the MAWP and execute a missed approach. The approach active annunciator and/or the receiver should be checked to **ensure the approach mode is active prior to the FAWP.**

8. Do not attempt to fly an approach unless the procedure is contained in the current, on-board navigation database and identified as "GPS" on the approach chart. The navigation database may contain information about nonoverlay approach procedures that is intended to be used to enhance position orientation, generally by providing a map, while flying these approaches using conventional NAVAIDs. This approach information should not be confused with a GPS overlay approach (see the receiver operating manual, AFM, or AFM Supplement for details on how to identify these procedures in the navigation database). Flying point to point on the approach does not assure compliance with the published approach procedure. The proper RAIM sensitivity will not be available and the CDI sensitivity will not automatically change to ±0.3 NM. Manually setting

CDI sensitivity does not automatically change the RAIM sensitivity on some receivers. Some existing nonprecision approach procedures cannot be coded for use with GPS and will not be available as overlays.

9. Pilots should pay particular attention to the exact operation of their GPS receivers for performing holding patterns and in the case of overlay approaches, operations such as procedure turns. These procedures may require manual intervention by the pilot to stop the sequencing of waypoints by the receiver and to resume automatic GPS navigation sequencing once the maneuver is complete. The same waypoint may appear in the route of flight more than once consecutively (e.g., IAWP, FAWP, MAHWP on a procedure turn). Care must be exercised to ensure that the receiver is sequenced to the appropriate waypoint for the segment of the procedure being flown, especially if one or more fly-overs are skipped (e.g., FAWP rather than IAWP if the procedure turn is not flown). The pilot may have to sequence past one or more fly-overs of the same waypoint in order to start GPS automatic sequencing at the proper place in the sequence of waypoints.

10. Incorrect inputs into the GPS receiver are especially critical during approaches. In some cases, an incorrect entry can cause the receiver to leave the approach mode.

11. A fix on an overlay approach identified by a DME fix will not be in the waypoint sequence on the GPS receiver unless there is a published name assigned to it. When a name is assigned, the along track to the waypoint may be zero rather than the DME stated on the approach chart. The pilot should be alert for this on any overlay procedure where the original approach used DME.

12. If a visual descent point (VDP) is published, it will not be included in the sequence of waypoints. Pilots are expected to use normal piloting techniques for beginning the visual descent, such as ATD.

13. Unnamed stepdown fixes in the final approach segment will not be coded in the waypoint sequence of the aircraft's navigation database and must be identified using ATD. Stepdown fixes in the final approach segment of RNAV (GPS) approaches are being named, in addition to being identified by ATD. However, since most GPS avionics do not accommodate waypoints between the FAF and MAP, even when the waypoint is named, the waypoints for these stepdown fixes may not appear in the sequence of waypoints in the navigation database. Pilots must continue to identify these stepdown fixes using ATD.

p. Missed Approach

1. **A GPS missed approach requires pilot action** to sequence the receiver past the MAWP to the missed approach portion of the procedure. The pilot must be thoroughly familiar with the activation procedure for the particular GPS receiver installed in the aircraft and **must initiate appropriate action after the MAWP.** Activating the missed approach prior to the MAWP will cause CDI sensitivity to immediately change to terminal (t1 NM) sensitivity and the

receiver will continue to navigate to the MAWP. The receiver will not sequence past the MAWP. Turns should not begin prior to the MAWP. If the missed approach is not activated, the GPS receiver will display an extension of the inbound final approach course and the ATD will increase from the MAWP until it is manually sequenced after crossing the MAWP.

2. Missed approach routings in which the first track is via a course rather than direct to the next waypoint require additional action by the pilot to set the course. Being familiar with all of the inputs required is especially critical during this phase of flight.

q. GPS Familiarization: Pilots should practice GPS approaches under visual meteorological conditions (VMC) until thoroughly proficient with all aspects of their equipment (receiver and installation) prior to attempting flight by IFR in instrument meteorological conditions (IMC). Some of the areas which the pilot should practice are:

1. Utilizing the receiver autonomous integrity monitoring (RAIM) prediction function;
2. Inserting a DP into the flight plan, including setting terminal CDI sensitivity, if required, and the conditions under which terminal RAIM is available for departure (some receivers are not DP or STAR capable);
3. Programming the destination airport;
4. Programming and flying the overlay approaches (especially procedure turns and arcs);
5. Changing to another approach after selecting an approach;
6. Programming and flying "direct" missed approaches;
7. Programming and flying "routed" missed approaches;
8. Entering, flying, and exiting holding patterns, particularly on overlay approaches with a second waypoint in the holding pattern;
9. Programming and flying a "route" from a holding pattern;
10. Programming and flying an approach with radar vectors to the intermediate segment;
11. Indication of the actions required for RAIM failure both before and after the FAWP; and
12. Programming a radial and distance from a VOR (often used in departure instructions).

1-1-21. Wide Area Augmentation System (WAAS)

a. General

1. The FAA developed the Wide Area Augmentation System (WAAS) to improve the accuracy, integrity and availability of GPS signals. WAAS will allow GPS to

be used, as the aviation navigation system, from takeoff through Category I precision approach when it is complete. WAAS is a Critical component of the FAA's strategic objective for a seamless navigation system for civil aviation, improving capacity and safety.

2. The International Civil Aviation Organization (ICAO) has defined Standards and Recommended Practices (SARPs) for satellite-based augmentation systems such as WAAS. Japan and Europe are building similar systems that are planned to be interoperable with WAAS:

    EGNOS, the European Geostationary Navigation Overlay System, and MSAS, the Japan Multifunctional Transport Satellite (MTSAT) Satellite-based Augmentation System. The merging of these systems will create a worldwide seamless navigation capability similar to GPS but with greater accuracy, availability and integrity.

3. Unlike traditional ground-based navigation aids, WAAS will cover a more extensive service area. Precisely surveyed wide-area ground reference stations (WRS) are linked to form the U.S. WAAS network. Signals from the GPS satellites are monitored by these WRSs to determine satellite clock and ephemeris corrections and to model the propagation effects of the ionosphere. Each station in the network relays the data to a wide-area master station (WMS) where the correction information is computed. A correction message is prepared and uplinked to a geostationary satellite (GEO) via a ground uplink station (GUS). The message is then broadcast on the same frequency as GPS (L1, 1575.42 MHz) to WAAS receivers within the broadcast coverage area of the WAAS GEO.

4. In addition to providing the correction signal, the WAAS GEO provides an additional pseudo range measurement to the aircraft receiver, improving the availability of GPS by providing, in effect, an additional satellite in view. The integrity of GPS is improved through real-time monitoring, and the accuracy is improved by providing differential corrections to reduce errors. The performance improvement is sufficient to enable approach procedures with GPS/WAAS glide paths (vertical guidance).

5. The FAA has completed 25 WRSs, 2 WMSs, 4 GUSs, and the required terrestrial communications to support the WAAS network. Prior to commissioning of the WAAS for public use, the FAA has been conducting a series of test and validation activities. Enhancements to the initial phase of WAAS will include additional master and reference stations, communications satellites, and transmission frequencies as needed.

6. GNSS navigation, including GPS and WAAS, is referenced to the WGS-84 coordinate system. It should only be used where the Aeronautical Information Publications (including electronic data and aeronautical charts) conform to WGS-84 or equivalent. Other countries civil aviation authorities may impose additional limitations on the use of their SBAS systems.

b. Instrument Approach Capabilities

1. A new class of approach procedures which provide vertical guidance, but which do not meet the ICAO Annex 10 requirements for precision approaches has been developed to support satellite navigation applications worldwide. These new procedures called Approach with Vertical Guidance (APV) are defined in ICAO Annex 6, and include approaches such as the LNAV/VNAV procedures presently being flown with barometric vertical navigation (Baro-VNAV). These approaches provide vertical guidance, but do not meet the more stringent standards of a precision approach. Properly certified WAAS receivers will be able to fly these LNAV/VNAV procedures using a WAAS electronic glide path, which eliminates the errors that can be introduced by using Barometric altimetry.

2. A new type of APV approach procedure, in addition to LNAV/VNAV, is being complemented to take advantage of the lateral precision provided by WAAS. This lateral precision, combined with an electronic glidepath allow the use of TERPS approach criteria very similar to that used for present precision approaches, with adjustments for the larger vertical containment limit. The resulting approach procedure minima, titled LPV, may have decision altitudes as low as 250 feet height above touchdown with visibility minimums as low as 1/2 mile, when the terrain and airport infrastructure support the lowest minima. LPV will be published in the RNAV (GPS) approach charts (see paragraph 5-4-5, Instrument Approach Procedure Charts).

3. WAAS initial operating capability provides a level of service that supports all 7 phases of flight, including LNAV, LNAV/VNAV, and LPV approaches. In the long term, WAAS will provide Category I precision approach services in conjunction with modernized GPS.

c. General Requirements

1. WAAS navigation equipment used must be approved in accordance with the requirements specified in Technical Standard Order (TSO) C145A or C146A and installed in accordance with Advisory Circular AC 20-138A, Airworthiness Approval of Global Navigation Satellite System (GNSS) Equipment.

2. GPS/WAAS operation must be conducted in accordance with the FAA-approved aircraft flight manual (AFM) and flight manual supplements. Flight manual supplements will state the level of approach procedure that the receiver supports. IFR approved WAAS receivers support all GPS only operations as long a lateral capability at the appropriate level is functional. WAAS monitors both GPS and WAAS satellites and provides integrity.

3. GPS/WAAS equipment is inherently capable of supporting oceanic and remote operations if the operator obtains a fault detection and exclusion (FDE) prediction program.

4. Air carrier and commercial operators must meet the appropriate provisions of their approved operations specifications.

5. Prior to GPS/WAAS IFR operation, the pilot must review appropriate Notices to Airmen (NOTAMs) and aeronautical information. The FAA will provide NOTAMs to advise pilots of the status of the WAAS. This will also include NOTAMs advising if the operational capability of GPS/WAAS equipment is degraded.

6. GPS/WAAS was developed to be used within SBAS GEO coverage (WAAS or other interoperable system) without the need for other radio navigation equipment appropriate to the route of flight to be flown. Outside the SBAS coverage or in the event of a WAAS failure, GPS/WAAS equipment reverts to GPS-only operation and satisfies the requirements for basic GPS equipment.

d. Flying Procedures with WAAS

1. WAAS receivers support all basic GPS approach functions and will provide additional capabilities. One of the major improvements is the ability to generate an electronic glide path, independent of ground equipment or barometric aiding. This eliminates several problems such as cold temperature effects, incorrect altimeter setting or lack of a local altimeter source and allows approach procedures to be built without the cost of installing ground stations at each airport. Some approach certified receivers will only support a glide path with performance similar to Baro-VNAV, and are authorized to fly the LNAV/VNAV line of minima on the RNAV (GPS) approach charts. Receivers with additional capability which support the performance requirements for precision approaches (including update rates and integrity limits) will be authorized to fly the LPV line of minima. The lateral integrity changes dramatically from the 0.3 NM (556 meter) limit for GPS, LNAV and LNAV/VNAV approach mode, to 40 meters for LPV. It also adds vertical integrity monitoring, which for LNAV/VNAV and LPV approaches bounds the vertical error to 50 meters.

2. When an approach procedure is selected and active, the receiver will notify the pilot of the most accurate level of service supported by the combination of the WAAS signal, the receiver, and the selected approach procedure. For example, if an approach is published with LPV minima and the receiver is only certified for LNAV/VNAV, the equipment would indicate "LPV not available - use LNAV/VNAV minima," even though the WAAS signal would support LPV. If flying an existing LNAV/VNAV procedure, the receiver will notify the pilot "LNAV/VNAV available" even if the receiver is certified for LPV and the WAAS signal supports LPV. If the WAAS signal does not support published minima lines which the receiver is certified to fly, the receiver will notify the pilot with a message such as "LPV not available - use the LNAV/VNAV minima." Once this notification has been given, the receiver will operate in this mode for the duration of the approach procedure. The receiver cannot change back to a more accurate level of service until the next time an approach is activated.

3. Another additional feature of WAAS receivers is the ability to exclude a bad GPS signal and continue operating normally. This is normally accomplished by the WAAS correction information. Outside WAAS coverage or when WAAS is not available, it is accomplished through a receiver algorithm called FDE. In most cases this operation will be invisible to the pilot since the receiver will continue to operate with other available satellites after excluding the "bad" signal. This capability increases the reliability of navigation.

4. Both lateral and vertical scaling for the LNAV/VNAV and LPV approach procedures are different than the linear scaling of basic GPS. When the complete published procedure is flown, ±1 NM linear scaling is provided until two (2) NM prior to the FAF, where the sensitivity increases to be similar to the angular scaling of an ILS. There are two differences in the WAAS scaling and the ILS: 1) on long final approach segments, the initial scaling will be ±0.3 NM to achieve equivalent performance to GPS (and better than ILS, which is less sensitive far from the runway); 2) close to the runway threshold, the scaling changes to linear instead of continuing to become more sensitive. The width of the final approach course is tailored so that it is usually 350 feet wide at the runway threshold, unlike ILS where the width varies with runway length. When the complete published procedure is not flown, and instead the aircraft needs to capture the extended final approach course similar to ILS, the vector to final (VTF) mode is used. Under VTF the scaling is linear at ±1 NM until the point where the angular splay reaches a width of ±1 NM regardless of the distance from the FAWP.

5. The WAAS scaling is also different than GPS TSO-C129 in the initial portion of the missed approach. Two differences occur here. First, the scaling abruptly changes from the approach scaling to the missed approach scaling, at approximately the departure end of the runway or when the pilot requests missed approach guidance rather than ramping as GPS does. Second, when the first leg of the missed approach is a Track to Fix (TF) leg aligned within 3° of the inbound course, the receiver will change to 0.3 nm linear sensitivity until the turn initiation point for the first waypoint in the missed approach procedure, at which time it will abruptly change to terminal (± nm) sensitivity. This allows the elimination of close in obstacles in the early part of the missed approach that nay cause the DA to be raised.

6. A new method has been added for selecting the final approach segment of an instrument approach. Along with the current method used by most receivers when the pilot selects the airport, the runway, the specific approach procedure and finally the IAF, there is also a channel number selection method. The pilot enters a unique 5-digit number provided on the approach chart, and the receiver recalls the matching final approach segment from the aircraft database. A list of information including the available IAFs is displayed and the pilot selects the appropriate IAF. The pilot should confirm that the correct final

approach segment was loaded by cross checking the approach ID, which is also provided on the approach chart.

7. The Along Track Distance (ATD) during the final approach segment of an LNAV procedure (with a minimum descent altitude) will be to the MAWP. On LNAV/VNAV and LPV approaches to a decision altitude, there is no missed approach waypoint so the along-track distance is displayed to a point normally located at the runway threshold. In most cases the MAWP for the LNAV approach is located on the runway threshold at the centerline, so these distances will be the same. This distance will always vary slightly from any ILS/DME that may be present, since the ILS/DME is located further down the runway. Initiation of the missed approach on the LNAV/VNAV and LPV approaches is still based on reaching the decision altitude without any of the items listed in 14 CFR Section 91.175 being visible, and must not be delayed until the ATD reaches zero. The WAAS receiver, unlike a GPS receiver, will automatically sequence past the MAWP if the missed approach procedure has been designed for RNAV. The pilot may also select missed approach prior to the MAWP, however, navigation will continue o the MAWP prior to waypoint sequencing taking place.

5-4-5. Instrument Approach Procedure Charts

a. 14 CFR Section 91.175(a), Instrument approaches to civil airports, requires the use of SIAPs prescribed for the airport in 14 CFR Part 97 unless otherwise authorized by the Administrator (including ATC). If there are military procedures published at a civil airport, aircraft operating under 14 CFR Section 91 must use the civil procedure(s). Civil procedures are defined with "FAA" in parenthesis; e.g., (FAA), at the top, center of the procedure chart. If a civil procedure does not exist to the required runway and a DOD procedure is published, civil aircraft may use the DOD procedure. DOD procedures are defined using the abbreviation of the applicable military service in parenthesis; e.g., (USAF), (USN), (USA). 14 CFR Section 91.175(g), Military airports, requires civil pilots flying into or out of military airports to comply with the IAPs and takeoff and landing minimums prescribed by the authority having jurisdiction at those airports. Unless an emergency exists, civil aircraft operating at military airports normally require advance authorization, commonly referred to as "Prior Permission Required" or "PPR." Information on obtaining a PPR for a particular military airport can be found in the Airport Facility Directory.

Note: *Civil aircraft may conduct practice VFR approaches using DOD instrument approach procedures when approved by the air traffic controller.*

1. IAPs (standard and special, civil and military) are based on joint civil and military criteria contained in the U.S. Standard for TERPS. The design of IAPs based on criteria contained in TERPS, takes into account the interre-

lationship between airports, facilities, and the surrounding environment, terrain, obstacles, noise sensitivity, etc. Appropriate altitudes, courses, headings, distances, and other limitations are specified and, once approved, the procedures are published and distributed by government and commercial cartographers as instrument approach charts.

2. Not all APs are published in chart form. Radar IAPs are established where requirements and facilities exist but they are printed in tabular form in appropriate U.S. Government Flight Information Publications.

3. Straight-in IAPs are identified by the navigational system providing the final approach guidance and the runway to which the approach is aligned (e.g. VOR RWY 13). Circling only approaches are identified by the navigational system providing final approach guidance and a letter (e.g., VOR A). More than one navigational system separated by a slash indicates that more than one type of equipment must be used to execute the final approach (e.g., VOR/DME RWY 31). More than one navigational system separated by the word "or" indicates either type of equipment may be used to execute the final approach (e.g., VOR or GPS RWY 15). In some cases, other types of navigation systems may be required to execute other portions of the approach (e.g., an NDB procedure turn to an ILS or an NDB in the missed approach). Pilots should ensure that the aircraft is equipped with the required NAVAID(s) in order to execute the approach, including the missed approach. The FAA has initiated a program to provide a new notation for LOC approaches when charted on an ILS approach requiring other navigational aids to fly the final approach course. The LOC minimums will be annotated with the NAVAID required e.g., "DME Required" or "RADAR Required." During the transition period, ILS approaches will still exist without the annotation. The naming of multiple approaches of the same type to the same runway is also changing. Multiple approaches with the same guidance will be annotated with an alphabetical suffix beginning at the end of the alphabet and working backwards for subsequent procedures (ILS Z RWY 28, ILS Y RWY 28, etc.). The existing annotations such as ILS 2 RWY 28 or Silver ILS RWY 28 will be phased out and replaced with the new designation. The Cat II and Cat III designations are used to differentiate between multiple ILSs to the same runway unless there are multiples of the same type. WAAS (LPV and LNAV/VNAV), and GPS (LNAV) approach procedures will be charted as RNAV (GPS) RWY (Number); e.g., RNAV (GPS) RWY 21. VOR/DME RNAV approaches will continue to be identified as VOR/DME RNAV RWY (Number); e.g., VOR/DME RNAV RWY 21. VOR/DME RNAV procedures which can be flown by GPS will be annotated with "or GPS" e.g., VOR/DME RNAV or GPS RWY 31.

4. Approach minimums are based on the local altimeter setting for that airport, unless annotated otherwise; e.g., Oklahoma City/Will Rogers World

approaches are based on having a Will Rogers World altimeter setting. When a different altimeter source is required, or more than one source is authorized, it will be annotated on the approach chart; e.g., use Sidney altimeter setting, if not received, use Scottsbluff altimeter setting. Approach minimums may be raised when a nonlocal altimeter source is authorized. When more than one altimeter source is authorized, and the minima are different, they will be shown by separate lines in the approach minima box or a note; e.g., use Manhattan altimeter setting; when not available use Salina altimeter setting and increase all MDAs 40 feet. When the altimeter must be obtained from a source other than air traffic a note will indicate the source; e.g., obtain local altimeter setting on CTAF. When the altimeter setting(s) on which the approach is based is not available, the approach is not authorized. Baro-VNAV must be flown using the local altimeter setting only. Where no local altimeter is available, the LNAV/VNAV line will still be published for use by WAAS receivers with a note that Baro-VNAV is not authorized. When a local and at least one other altimeter setting source is authorized and the local altimeter is not available Baro-VNAV is not authorized; however, the LNAV/VNAV minima can still be used by WAAS receivers using the alternate altimeter setting source.

5. A pilot adhering to the altitudes, flight paths, and weather minimums depicted on the IAP chart or vectors and altitudes issued by the radar controller, is assured of terrain and obstruction clearance and runway or airport alignment during approach for landing.

6. IAPs are designed to provide an IFR descent from the en route environment to a point where a safe landing can be made. They are prescribed and approved by appropriate civil or military authority to ensure a safe descent during instrument flight conditions at a specific airport. It is important that pilots understand these procedures and their use prior to attempting to fly instrument approaches.

7. TERPS criteria are provided for the following types of instrument approach procedures:

   (a) Precision Approach (PA). An instrument approach based on a navigation system that provides course and glidepath deviation information meeting the precision standards of ICAO Annex 10. For example, PAR, ILS, and GLS are precision approaches.

   (b) Approach with Vertical Guidance (APV). An instrument approach based on a navigation system that is not required to meet the precision approach standards of ICAO Annex 10 but provides course and glidepath deviation information. For example, Baro-VNAV, LDA with glidepath, LNAV/VNAV and LPV are APV approaches.

   (c) Nonprecision Approach (NPA). An instrument approach based on a navigation system which provides course deviation information, but no glidepath deviation information. For example, VOR, NDB and LNAV.

As noted in subparagraph h, Vertical Descent Angle (VDA) on Non-precision Approaches, some approach procedures may provide a Vertical Descent Angle as an aid in flying a stabilized approach, without requiring its use in order to fly the procedure. This does not make the approach an APV procedure, since it must still be flown to an MDA and has not been evaluated with a glidepath.

b. The method used to depict prescribed altitudes on instrument approach charts differs according to techniques employed by different chart publishers. Prescribed altitudes may be depicted in three different configurations: minimum, maximum, and mandatory. The U.S. Government distributes charts produced by National Imagery and Mapping Agency (NIMA) and FAA. Altitudes are depicted on these charts in the profile view with underscore, overscore, or both to identify them as minimum, maximum, or mandatory.

1. Minimum altitude will be depicted with the altitude value underscored. Aircraft are required to maintain altitude at or above the depicted value.

2. Maximum altitude will be depicted with the altitude value overscored. Aircraft are required to maintain altitude at or below the depicted value.

3. Mandatory altitude will be depicted with the altitude value both underscored and overscored. Aircraft are required to maintain altitude at the depicted value.

Note: *The underscore and overscore to identify mandatory altitudes and the overscore to identify maximum altitudes are used almost exclusively by NIMA for military charts. With very few exceptions, civil approach charts produced by FAA utilize only the underscore to identify minimum altitudes. Pilots are cautioned to adhere to altitudes as prescribed because, in certain instances, they may be used as the basis for vertical separation of aircraft by ATC. When a depicted altitude is specified in the ATC clearance, that altitude becomes mandatory as defined above.*

c. Minimum Safe/Sector Altitudes (MSA) are published for emergency use on UP charts. For conventional navigation systems, the MSA is normally based on the primary omnidirectional facility on which the UP is predicated. The MSA depiction on the approach chart contains the facility identifier of the NAVAID used to determine the MSA altitudes. For RNAV approaches, the MSA is based on the runway waypoint (RWY WP) for straight-in approaches, or the airport waypoint (APT WP) for circling approaches. For GPS approaches, the MSA center will be the missed approach waypoint (MAW). MSAs are expressed in feet above mean sea level and normally have a 25 NM radius; however, this radius may be expanded to 30 NM if necessary to encompass the airport landing surfaces. Ideally, a single sector altitude is established and depicted on the plan view of approach charts; however, when necessary to obtain relief from obstructions, the area may be further sectored and as many as four MSAs established. When established, sectors may be no less than 90° in spread. MSAs provide 1,000 feet clearance over all obstructions but do not necessarily assure acceptable navigation signal coverage.

d. Terminal Arrival Area (TAA)

1. The objective of the TAA is to provide a seamless transition from the en route structure to the terminal environment for arriving aircraft equipped with Flight Management System (FMS) and/or Global Positioning System (GPS) navigational equipment. The underlying instrument approach procedure is an area navigation (RNAV) procedure described in this section. The TAA provides the pilot and air traffic controller with a very efficient method for routing traffic into the terminal environment with little required air traffic control interface, and with minimum altitudes depicted that provide standard obstacle clearance compatible with the instrument procedure associated with it. The TAA will not be found on all RNAV procedures, particularly in areas of heavy concentration of air traffic. When the TAA is published, it replaces the MSA for that approach procedure. See Fig. 5-4-9 for a depiction of a RNAV approach chart with a TAA.

2. The RNAV procedure underlying the TAA will be the "T" design (also called the "Basic T"), or a modification of the "T." The "T" design incorporates from one to three IAFs; an intermediate fix (IF) that serves as a dual purpose IF (IAF); a final approach fix (FAF), and a missed approach point (MAP) usually located at the runway threshold. The three IAFs are normally aligned in a straight line perpendicular to the intermediate course, which is an extension of the final course leading to the runway, forming a "T." The initial segment is normally from 3-6 NM in length; the intermediate 5-7 NM, and the final segment 5 NM. Specific segment length may be varied to accommodate specific aircraft categories for which the procedure is designed. However, the published segment lengths will reflect the highest category of aircraft normally expected to use the procedure.

   (a) A standard racetrack holding pattern may be provided at the center IAF, and if present may be necessary for course reversal and for altitude adjustment for entry into the procedure. In the latter case, the pattern provides an extended distance for the descent required by the procedure. Depiction of this pattern in U.S. Government publications will utilize the "hold-in-lieu-of-PT" holding pattern symbol.

   (b) The published procedure will be annotated to indicate when the course reversal is not necessary when flying within a particular TAA area; e.g., "NoPT." Otherwise, the pilot is expected to execute the course reversal under the provisions of 14 CFR Section 91.175. The pilot may elect to use the course reversal pattern when it is not required by the procedure, but must inform air traffic control and receive clearance to do so. (See Fig. 5-4-1 and Fig. 5-4-2.)

3. The "T" design may be modified by the procedure designers where required by terrain or air traffic control considerations. For instance, the

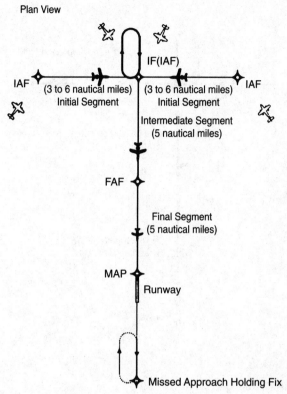

Figure 5-4-1 Basic "T" design.

"T" design may appear more like a regularly or irregularly shaped "Y", or may even have one or both outboard IAFs eliminated resulting in an upside down "L" or an "I" configuration. (See Fig. 5-4-3 and Fig. 5-4-10). Further, the leg lengths associated with the outboard. IAFs may differ. (See Fig. 5-4-5 and Fig. 5-4-6.)

4. Another modification of the "T" design may be found at airports with parallel runway configurations. Each parallel runway may be served by its own "T" IAF, IF (LAY), and FAF combination, resulting in parallel final approach courses. (See Fig. 5-4-4.) Common IAFs may serve both runways; however, only the intermediate and final approach segments for the landing runway will be shown on the approach chart. (See Fig. 5-4-5 and Fig. 5-4-6.)

5. The standard TAA consists of three areas defined by the extension of the IAF legs and the intermediate segment course. These areas are called the straight-in, left-base, and right-base areas. (See Fig. 5-4-7.) TAA area lateral boundaries are identified by magnetic courses TO the IF (IAF). The

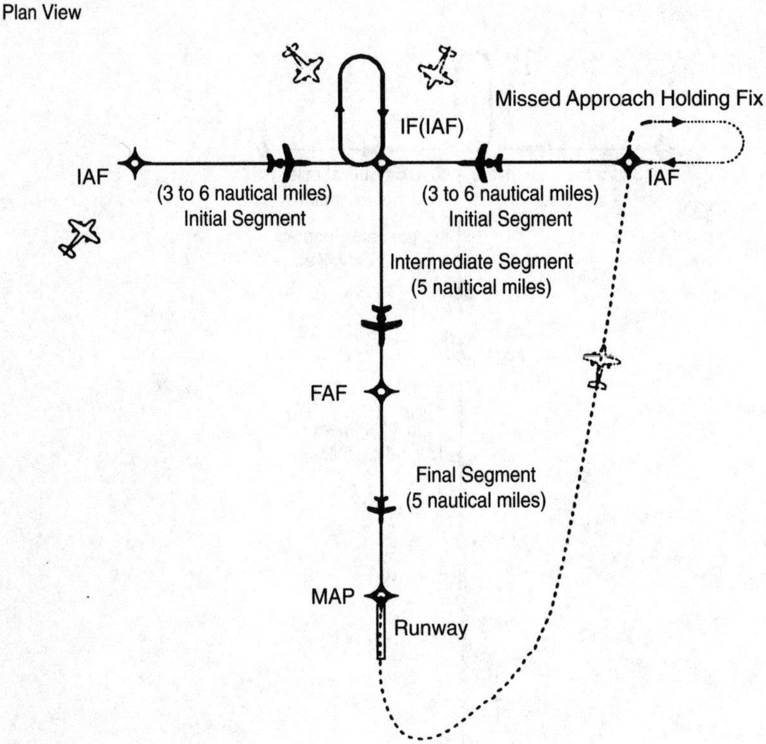

Plan View

Figure 5-4-2 Basic "T" design.

straight-in area can be further divided into pie-shaped sectors with the boundaries identified by magnetic courses TO the IF (IAF), and may contain stepdown sections defined by arcs based on RNAV distances (DME or ATD) from the IF (IAF). The right/left-base areas can only be subdivided using arcs based on RNAV distances from the IAFs for those areas. Minimum MSL altitudes are charted within each of these defined areas/subdivisions that provide at least 1,000 feet of obstacle clearance, or more as necessary in mountainous areas.

(a) Prior to arriving at the TAA boundary, the pilot can determine which area of the TAA the aircraft will enter by selecting the IF (IAF) to determine the magnetic bearing TO the center IF (IAF). That bearing should then be compared with the published bearings that define the lateral boundaries of the TAA areas. Using the end IAFs may give a false indication of which area the aircraft will enter. This is critical when approaching the TAA near the extended boundary between the

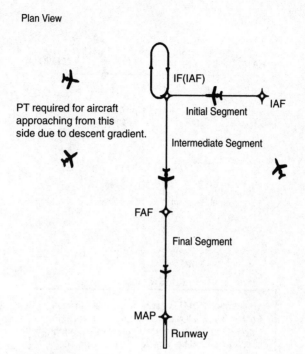

Plan View

IF(IAF)

IAF

PT required for aircraft approaching from this side due to descent gradient.

Initial Segment

Intermediate Segment

FAF

Final Segment

MAP

Runway

Figure 5-4-3 Modified basic "T".

left and right-base areas, especially where these areas contain different minimum altitude requirements.

(b) Pilots entering the TAA and cleared by air traffic control, are expected to proceed directly to the IAF associated with that area of the TAA at the altitude depicted, unless otherwise cleared by air traffic control. Pilots entering the TAA with two-way radio communications failure (14 CFR Section 91.185, IFR Operations: Two-way Radio Communications Failure), must maintain the highest altitude prescribed by Section 91.185 (c) (2) until arriving at the appropriate IAF.

(c) Depiction of the TAA on U.S. Government charts will be through the use of icons located in the plan view outside the depiction of the actual approach procedure. (See Fig. 5-4-9.) Use of icons is necessary to avoid obscuring any portion of the "T" procedure (altitudes, courses, minimum altitudes, etc.). The icon for each TAA area will be located and oriented on the plan view with respect to the direction of arrival to the approach procedure, and will show all TAA minimum altitudes and sector/radius subdivisions for that area. The IAF for each area of the TAA is included on the icon where it appears on the approach, to

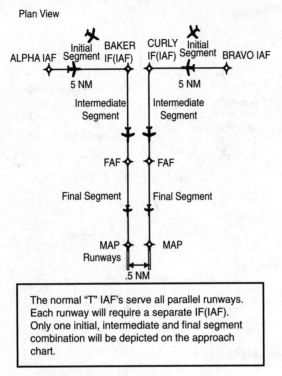

Figure 5-4-4 Modified "T" approach to parallel runways.

help the pilot orient the icon to the approach procedure. The IAF name and the distance of the TAA area boundary from the IAF are included on the outside arc of the TAA area icon. Examples here are shown with the TAA around the approach to aid pilots in visualizing how the TAA corresponds to the approach and should not be confused with the actual approach chart depiction.

(d) Each waypoint on the "T", except the missed approach waypoint, is assigned a pronounceable 5-character name used in air traffic control communications, and which is found in the RNAV databases for the procedure. The missed approach waypoint is assigned a pronounceable name when it is not located at the runway threshold.

6. Once cleared to fly the TAA, pilots are expected to obey minimum altitudes depicted within the TAA icons, unless instructed otherwise by air traffic control. In Fig. 5-4-8, pilots within the left- or right-base areas are expected to maintain a minimum altitude of 6,000 feet until within 17 NM of the associated IAF. After crossing the 17 NM arc, descent is authorized to the lower charted altitudes. Pilots approaching from the northwest are

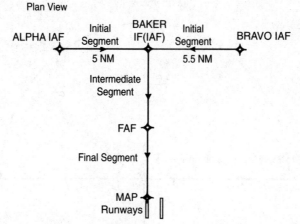

Figure 5-4-5 "T" approach with common IAFs to parallel runways.

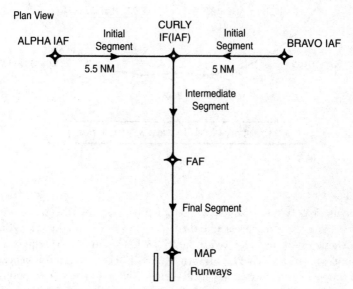

Figure 5-4-6 "T" approach with common IAFs to parallel runways.

expected to maintain a minimum altitude of 6,000 feet, and when within 22 NM of the IF (IAF), descend to a minimum altitude of 2,000 feet MSL until reaching the IF (IAF).

7. Just as the underlying "T" approach procedure may be modified in shape, the TAA may contain modifications to the defined area shapes and sizes. Some areas may even be eliminated, with other areas expanded as needed.

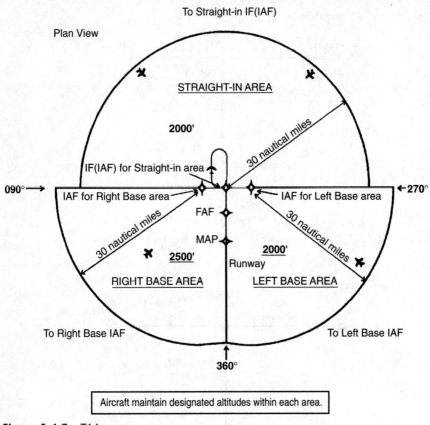

Plan View

To Straight-in IF(IAF)

STRAIGHT-IN AREA

2000'

30 nautical miles

IF(IAF) for Straight-in area

090° →   ← 270°

IAF for Right Base area   IAF for Left Base area

30 nautical miles

FAF

MAP

2500'   2000'

Runway

RIGHT BASE AREA   LEFT BASE AREA

30 nautical miles

To Right Base IAF   To Left Base IAF

360°

Aircraft maintain designated altitudes within each area.

Figure 5-4-7   TAA area.

Figure 5–4–10 is an example of a design limitation where a course reversal is necessary when approaching the IF (IAF) from certain directions due to the amount of turn required at the IF (IAF). Design criteria require a course reversal whenever this turn exceeds 120 degrees. In this generalized example, pilots approaching on a bearing TO the IF (IAF) from 300° clockwise through 060° are expected to execute a course reversal. The term "NoPT" will be annotated on the boundary of the TAA icon for the other portion of the TAA.

8.   Figure 5–4–11 depicts another TAA modification that pilots may encounter. In this generalized example, the right-base area has been eliminated. Pilots operating within the TAA between 360° clockwise to 060° bearing TO the IF (IAF) are expected to execute the course reversal in order to properly align the aircraft for entry onto the intermediate segment. Aircraft operating in all other areas from 060° clockwise to 360° bearing TO the IF (LAY)

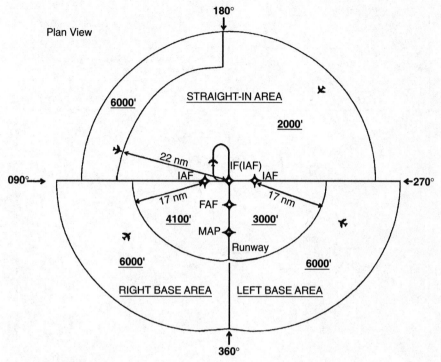

Figure 5-4-8 Sectored TAA areas.

need not perform the course reversal, and the term "NoPT" will be annotated on the TAA boundary of the icon in these areas. TAAs are no longer
being produced with sections removed; however, some may still exist on
previously published procedures.

9. When an airway does not cross the lateral TAA boundaries, a feeder route
    will be established to provide a transition from the en route structure to the
    appropriate IAF. Each feeder route will terminate at the TAA boundary, and
    will be aligned along a path pointing to the associated IAF. Pilots should
    descend to the TAA altitude after crossing the TAA boundary and cleared
    by air traffic control. (See Fig. 5-4-12.)

e. Minimum Vectoring Altitudes (MVAs) are established for use by ATC
when radar ATC is exercised. MVA charts are prepared by air traffic facilities
at locations where there are numerous different minimum IFR altitudes. Each
MVA chart has sectors large enough to accommodate vectoring of aircraft
within the sector at the MVA. Each sector boundary is at least 3 miles from
the obstruction determining the MVA. To avoid a large sector with an excessively high MVA due to an isolated prominent obstruction, the obstruction
may be enclosed in a buffer area whose boundaries are at least 3 miles from

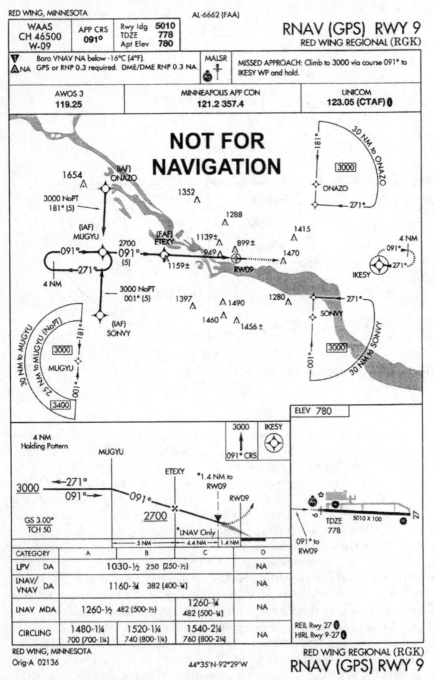

Figure 5-4-9 RNAV (GPS) approach chart. (*Note:* This chart has been modified to depict new concepts and may not reflect actual approach minima.)

Plan View

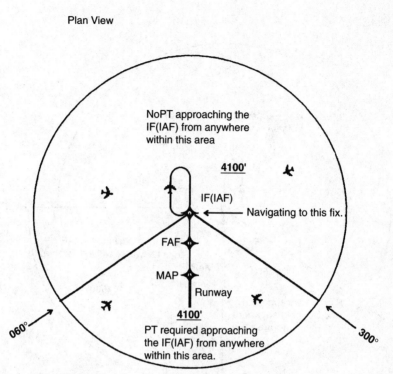

Figure 5-4-10 TAA with left and right base areas eliminated.

the obstruction. This is done to facilitate vectoring around the obstruction. (See Fig. 5-4-13.)

1. The minimum vectoring altitude in each sector provides 1,000 feet above the highest obstacle in nonmountainous areas and 2,000 feet above the highest obstacle in designated mountainous areas. Where lower MVAs are required in designated mountainous areas to achieve compatibility with terminal routes or to permit vectoring to an IAP, 1,000 feet of obstacle clearance may be authorized with the use of Airport Surveillance Radar (ASR). The minimum vectoring altitude will provide at least 300 feet above the floor of controlled airspace.

Note: *OROCA is an off-route altitude which provides obstruction clearance with a 1,000 foot buffer in nonmountainous terrain areas and a 2,000 foot buffer in designated mountainous areas within the U.S. This altitude may not provide signal coverage from ground-based navigational aids, air traffic control radar, or communications coverage.*

2. Because of differences in the areas considered for MVA, and those applied to other minimum altitudes, and the ability to isolate specific obstacles,

Plan View

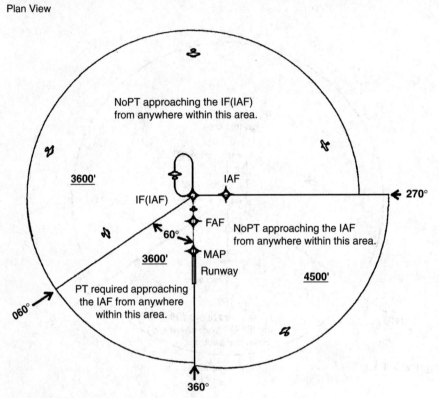

Figure 5-4-11   TAA with right base eliminated.

some MVAs may be lower than the nonradar Minimum En Route Altitudes (MEAs), Minimum Obstruction Clearance Altitudes (MOCAs) or other minimum altitudes depicted on charts for a given location. While being radar vectored, IFR altitude assignments by ATC will be at or above MVA.

f. **Visual Descent Points (VDPs)** are being incorporated in nonprecision approach procedures. The VDP is a defined point on the final approach course of a nonprecision straight-in approach procedure from which normal descent from the MDA to the runway touchdown point may be commenced, provided visual reference required by 14 CFR Section 91.175(c)(3) is established. The VDP will normally be identified by DME on VOR and LOC procedures and by along track distance to the next waypoint for RNAV procedures. The VDP is identified on the profile view of the approach chart by the symbol: **V.**

1. VDPs are intended to provide additional guidance where they are implemented. No special technique is required to fly a procedure with a VDP. The pilot should not descend below the MDA prior to reaching the VDP and acquiring the necessary visual reference.

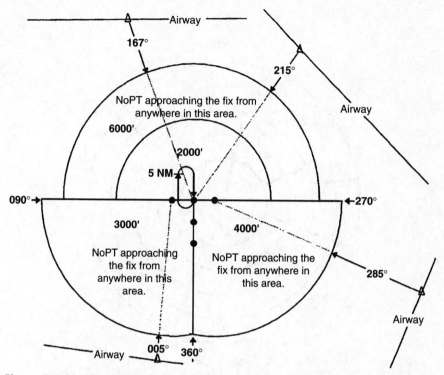

Figure 5-4-12   Examples of a TAA with feeders from an airway.

2. Pilots not equipped to receive the VDP should fly the approach procedure as though no VDP had been provided.

g. **Visual Portion of the Final Segment.** Instrument procedures designers perform a visual area obstruction evaluation off the approach end of each runway authorized for instrument landing, straight-in, or circling. Restrictions to instrument operations are imposed if penetrations of the obstruction clearance surfaces exist. These restrictions vary based on the severity of the penetrations, and may include increasing required visibility, denying VDPs and prohibiting night instrument operations to the runway.

h. **Vertical Descent Angle (VDA) on Nonprecision Approaches.** The FAA intends to eventually publish VDAs on all nonprecision approaches. Published along with the VDA is the threshold crossing height (TCH); i.e., the height of the descent angle above the landing threshold. The descent angle describes a computed path from the final approach fix (FAF) and altitude to the runway threshold at the published TCH. The optimum descent angle is 3.00 degrees; and whenever possible the approach will be designed to accommodate this angle.

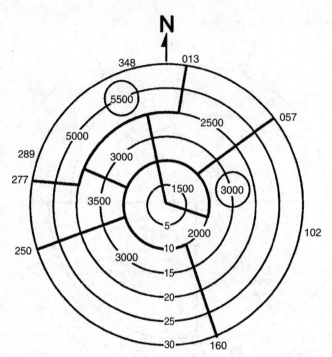

Figure 5-4-13 Minimum vectoring altitude charts.

1. The VDA provides the pilot with information not previously available on nonprecision approaches. It provides the means for the pilot to establish a stabilized approach descent from the FAF or stepdown fix to the TCH. Stabilized descent along this path is a key factor in the reduction of controlled flight into terrain (CFIT) incidents. Pilots can use the published angle and estimated/actual groundspeed to find a target rate of descent from a rate of descent table published in the back of the U.S. Terminal Procedures Publication.

2. Normally, the VDA will first appear on the nonprecision approach chart as the procedure is amended through the normal process. However, in some cases, pilots may see this data provided via a D-NOTAM.

Example

GPS RWY 9L, AMDT 2...
ADD: AWZAC WP TO RW09L: 2.96 DEGREES, TCH 50.
THIS IS GPS RWY 9L, AMDT 2A

Translated, this means that the currently published GPS RWY 9L procedure, Amendment 2, is changed by the addition of a 2.96-degree descent angle from

AWZAC WP to a point 50 feet above the RWY 9L threshold. This constitutes Amendment 2A to the **published procedure.**

3. Pilots should be aware that **the published angle is for information only—** it is strictly advisory in nature. There is no implicit additional obstacle protection below the MDA. Pilots must still respect the published minimum descent altitude (MDA) unless the visual cues stated in 14 CFR Section 91.175 are present. In rare cases, the published procedure descent angle will not coincide with the Visual Glide Slope Indicator (VGSI); VAST or PAPI. In these cases, the procedure will be annotated: "VGSI and descent angle not coincident."

i. **Pilot Operational Considerations When Flying Nonprecision Approaches.** The missed approach point (MAP) on a nonprecision approach is not designed with any consideration to where the aircraft must begin descent to execute a safe landing. It is developed based on terrain, obstructions, NAVAID location and possibly air traffic considerations. Because the MAP may be located anywhere from well prior to the runway threshold to past the opposite end of the runway, the descent from the Minimum Descent Altitude (MDA) to the runway threshold cannot be determined based on the MAP location. Descent from MDA at the MAP when the MAP is located close to the threshold would require an excessively steep descent gradient to land in the normal touchdown zone. Any turn from the final approach course to the runway heading may also be a factor in when to begin the descent.

1. Pilots are cautioned that descent to a straight-in landing from the MDA at the MAP may be inadvisable or impossible, on a nonprecision approach, even if current weather conditions meet the published ceiling and visibility. Aircraft speed, height above the runway, descent rate, amount of turn and runway length are some of the factors which must be considered by the pilot to determine if a landing can be accomplished.

2. Visual descent points (VDPs) provide pilots with a reference for the optimal location to begin descent from the MDA, based on the designed vertical descent angle (VDA) for the approach procedure, assuming required visual references are available. Approaches without VDPs have not been assessed for terrain clearance below the MDA, and may not provide a clear vertical path to the runway at the normally expected descent angle. Therefore, pilots must be especially vigilant when descending below the MDA at locations without VDPs. This does not necessarily prevent flying the normal angle; it only means that obstacle clearance in the visual segment could be less and greater care should be exercised in looking for obstacles in the visual segment. Use of visual glide slope indicator (VGSI) systems can aid the pilot in determining if the aircraft is in a position to make the descent from the MDA. However, when the visibility is close to minimums, the VGSI may not be visible at the start descent point for a "normal" glide path, due to its location down the runway.

3. Accordingly, pilots are advised to carefully review approach procedures, prior to initiating the approach, to identify the optimum position(s), and any unacceptable positions, from which a descent to landing can be initiated (in accordance with 14 CFR Section 91.175(c)).

j. **Area Navigation (RNAV) Instrument Approach Charts.** Reliance on RNAV systems for instrument approach operations is becoming more commonplace as new systems such as GPS, Wide Area Augmentation System (WAAS) and Local Area Augmentation System (LAAS) are developed and deployed. In order to foster and support full integration of RNAV into the National Airspace System (NAS), the FAA has developed a new charting format for IAPs. (See Fig. 5-4-9.) This format avoids unnecessary duplication and proliferation of instrument approach charts. The approach minimums for unaugmented GPS (the present GPS approaches) and WAAS augmented GPS will be published on the same approach chart. The approach chart will be titled "RNAV (GPS) RWY XX." The first new RNAV approach charts appeared as stand alone "GPS" procedures, prior to WAAS becoming operational, with only a LNAV minima line. The follow-on charts contained as many as four lines of approach minimums: GLS (Global Navigation Satellite System [GNSS] Landing System); LNAV/VNAV (lateral navigation/vertical navigation); LNAV; and CIRCLING. GLS was a placeholder for WAAS and LAAS when they became available and is marked N/A. LNAV/VNAV is a new type of instrument approach called APV, with lateral and vertical navigation. The vertical portion can be flown by approach certified Baro-VNAV and by WAAS electronic VNAV as well. A new line will be added to these charts titled LPV. This will replace the GLS N/A line. The decision of whether the WAAS and LAAS precision lines of minima will remain on this chart or be moved to a precision only chart will be determined later. RNAV procedures which incorporate a final approach stepdown fix may be published without vertical navigation, on a separate chart, titled RNAV (GPS) RWY XX and then Z, Y, X, etc., as indicated in subparagraph 5-4-5a3. During a transition period until all GPS procedures are retitled both "RNAV (GPS)" and "GPS" approach charts and formats will be published. ATC clearance for the RNAV procedure authorizes a properly certified pilot to utilize any landing minimums for which the aircraft is certified. The RNAV chart includes formatted information required for quick pilot or flight crew reference located at the top of the chart. This portion of the chart, developed based on a study by the Department of Transportation, Volpe National Transportation Systems Center, is commonly referred to as the pilot briefing.

1. The minima lines are:

(a) GLS. "GLS" is the acronym for GNSS Landing System; GNSS is the acronym for Global Navigation Satellite System. The minimums line labeled GLS will accommodate aircraft equipped with precision approach certified WAAS receivers operating to their fullest capability or LAAS and may support precision (GLS) approach minimums as low as 200-foot height above touchdown (HAT) and 1/2 statute mile (SM) visibility. This line has been published as N/A as a place holder for the minima when published. The WAAS and LAAS precision minima may be moved to a different chart in the future.

(b) LPV identifies APV minimums with electronic lateral and vertical guidance. The lateral guidance is equivalent to localizer, and the protected area is considerably smaller than the protected area for the present LNAV and LNAV/VNAV lateral protection. This minima line can be flown by aircraft with a statement in the Aircraft Flight Manual that the installed equipment supports LPV approach procedures. This includes Class 3 and 4 TSO-C146 WAAS equipment. Depending on the location of the obstacles around an airport this may result in lower minima than a LNAV/VNAV procedure. The vertical performance accommodates equipment with integrity limits larger than required for precision, but with precision approach 5 Hz update rates to help control flight technical error. The minima will be published as a DA. This procedure takes advantage of the precise lateral guidance available from the WAAS system at initial operational capability, allowing lower minimums at a large number of airports while the vertical integrity limits of WAAS are being further improved. This minima can also be used as a fail down mode if the WAAS vertical integrity does not meet the requirement for a precision approach.

(c) LNAV/VNAV identifies APV minimums developed to accommodate an RNAV LAP with vertical guidance, usually provided by approach certified Baro-VNAV, but with lateral and vertical integrity limits larger than a precision approach or LPV. RNAV stands for Lateral Navigation; VNAV stands for Vertical Navigation. This minima line can be flown by aircraft with a statement in the Aircraft Flight Manual that the installed equipment supports GPS approaches and has an approach-approved barometric VNAV, or if the aircraft has been demonstrated to support LNAV/VNAV approaches. This includes Class 2, 3 and 4 TSO-C146 WAAS equipment. Aircraft using LNAV/VNAV minimums will descend to landing via an internally generated descent path based on satellite or other approach approved VNAV systems. WAAS equipment may revert to this mode of operation when the signal does not support precision or LPV integrity. Since electronic vertical guidance is provided, the minima will be published as a DA. Other navigation systems may be specifically authorized to use this line of minima, see Section A, Terms/Landing Minima Data, of the U.S. Terminal Procedures books.

(d) LNAV. This minima is for lateral navigation only, and the approach minimum altitude will be published as a minimum descent altitude (MDA). RNAV provides the same level of service as the present GPS stand alone approaches. RNAV minimums support the following navigation systems: WAAS, when the navigation solution will not support vertical navigation; and, GPS navigation systems which are presently authorized to conduct

GPS approaches. Existing GPS approaches continue to be converted to the RNAV (GPS) format as they are revised or reviewed.

Note: *GPS receivers approved for approach operations in accordance with: AC 20-138, Airworthiness Approval of Global Positioning System (GPS) Navigation Equipment for Use as a VFR and IFR Supplemental Navigation System, for stand-alone Technical Standard Order (TSO) TSO-C129 Class A(1) systems; or AC 20-130A, Airworthiness Approval of Navigation or Flight Management Systems Integrating Multiple Navigation Sensors, for GPS as part of a multi-sensor system, qualify for this minima. WAAS navigation equipment must be approved in accordance with the requirements specified in TSO-C145 or TSO-C146 and installed in accordance with Advisory Circular AC 20-138A, Airworthiness Approval of Global Navigation Satellite System (GNSS) Equipment.*

2. Other systems may be authorized to utilize these approaches. See the description in Section A of the U.S. Terminal Procedures books for details. These systems may include aircraft equipped with an FMS that can file /E or /F. Operational approval must also be obtained for Baro-VNAV systems to operate to the LNAV/VNAV minimums. Baro-VNAV may not be authorized on some approaches due to other factors, such as no local altimeter source being available. Baro-VNAV is not authorized on LPV procedures. Pilots are directed to their local Flight Standards District Office (FSDO) for additional information.

Note: *RNAV and Baro-VNAV systems must have a manufacturer supplied electronic database which shall include the waypoints, altitudes, and vertical data for the procedure to be flown. The system shall also be able to extract the procedure in its entirety, not just as a manually entered series of waypoints.*

3. Required Navigation Performance (RNP)
   (a) Pilots are advised to refer to the "TERMS/LANDING MINIMUMS DATA" (Section A) of the U.S. Government Terminal Procedures books for aircraft approach eligibility requirements by specific RNP level requirements.
   (b) Some aircraft have RNP approval in their AFM without a GPS sensor. The lowest level of sensors that the FAA will support for RNP service is DME/DME. However, necessary DME signal may not be available at the airport of intended operations. For those locations having an RNAV chart published with LNAV/VNAV minimums, a procedure note may be provided such as "DME/DME RNP-0.3 NA." This means that RNP aircraft dependent on DME/DME to achieve RNP-0.3 are not authorized to conduct this approach. Where DME facility availability is a factor, the note may read "DME/DME RNP-0.3 Authorized; ABC and XYZ Required." This means that ABC and XYZ facilities have been determined by flight inspection to be required in the navigation solution to assure RNP-0.3. VOR/DME updating must not be used for approach procedures.

4. Chart Terminology
   (a) Decision Altitude (DA) replaces the familiar term Decision Height (DH). DA conforms to the international convention where altitudes relate to MSL and heights relate to AGL. DA will eventually be published for other types of instrument approach procedures with vertical guidance, as well. DA indicates to the pilot that the published descent profile is flown to the DA (MSL), where a missed approach will be initiated if visual references for landing are not established. Obstacle clearance is provided to allow a momentary descent below DA while transitioning from the final approach to the missed approach. The aircraft is expected to follow the missed instructions while continuing along the published final approach course to at least the published runway threshold waypoint or MAP (if not at the threshold) before executing any turns.
   (b) Minimum Descent Altitude (MDA) has been in use for many years, and will continue to be used for the RNAV only and circling procedures.
   (c) Threshold Crossing Height (TCH) has been traditionally used in "precision" approaches as the height of the glide slope above threshold. With publication of LNAV/VNAV minimums and LNAV descent angles, including graphically depicted descent profiles, TCH also applies to the height of the "descent angle," or glidepath, at the threshold. Unless otherwise required for larger type aircraft which may be using the IAP, the typical TCH is 30 to 50 feet.
5. The MINIMA FORMAT will also change slightly.
   (a) Each line of minima on the LNAV IAP is titled to reflect the level of service available; e.g., GLS, LPV, LNAV/VNAV, and LNAV. CIRCLING minima will also be provided.
   (b) The minima title box indicates the nature of the minimum altitude for the IAP. For example:
       (1) DA will be published next to the minima line title for minimums supporting vertical guidance such as for GLS, LPV or LNAV/VNAV.
       (2) MDA will be published where the minima line was designed to support aircraft with only lateral guidance available, such as LNAV. Descent below the MDA, including during the missed approach, is not authorized unless the visual conditions stated in 14 CFR Section 91.175 exist.
       (3) Where two or more systems, such as LPV and LNAV/VNAV, share the same minima, each line of minima will be displayed separately.
6. Chart Symbology changed slightly to include:
   (a) Descent Profile. The published descent profile and a graphical depiction of the vertical path to the runway will be shown. Graphical depiction of the LNAV vertical guidance will differ from the traditional depiction of an ILS glide slope (feather) through the use of a shorter vertical track beginning at the decision altitude.

(1) It is FAA policy to design IAPs with minimum altitudes established at fixes/waypoints to achieve optimum stabilized (constant rate) descents within each procedure segment. This design can enhance the safety of the operations and contribute toward reduction in the occurrence of controlled flight into terrain (CFIT) accidents. Additionally, the National Transportation Safety Board (NTSB) recently emphasized that pilots could benefit from publication of the appropriate LAP descent angle for a stabilized descent on final approach. The LNAV IAP format includes the descent angle to the hundredth of a degree; e.g., 3.00 degrees. The angle will be provided in the graphically depicted descent profile.

(2) The stabilized approach may be performed by reference to vertical navigation information provided by WAAS or LNAV/VNAV systems; or for LNAV-only systems, by the pilot determining the appropriate aircraft attitude/groundspeed combination to attain a constant rate descent which best emulates the published angle. To aid the pilot, U.S. Government Terminal Procedures Publication charts publish an expanded Rate of Descent Table on the inside of the back hard cover for use in planning and executing precision descents under known or approximate groundspeed conditions.

(b) Visual Descent Point (VDP). A VDP will be published on most LNAV IAPs. <u>VDPs apply only to aircraft utilizing LNAV minima</u>, not LPV or LNAV/VNAV minimums.

(c) Missed Approach Symbology. In order to make missed approach guidance more readily understood, a method has been developed to display missed approach guidance in the profile view through the use of quick reference icons. Due to limited space in the profile area, only four or fewer icons can be shown. However, the icons may not provide representation of the entire missed approach procedure. The entire set of textual missed approach instructions are provided at the top of the approach chart in the pilot briefing. (See Fig. 5-4-9).

(d) Waypoints. All LNAV or GPS stand-alone IAPs are flown using data pertaining to the particular IAP obtained from an onboard database, including the sequence of all WPs used for the approach and missed approach, except that step down waypoints may not be included in some TSO C-129 receiver databases. Included in the database, in most receivers, is coding that informs the navigation system of which WPs are fly-over (FO) or fly-by (FB). The navigation system may provide guidance appropriately - including leading the turn prior to a fly-by WP; or causing overflight of a fly-over WP. Where the navigation system does not provide such guidance, the pilot must accomplish the turn lead or waypoint overflight manually. Chart symbology for the FB WP provides pilot awareness of expected actions. Refer to the legend of the U.S. Terminal Procedures books.

(e) TAAs are described in paragraph 5-4-5d, Terminal Arrival Area (TAA). When published, the LNAV chart depicts the TAA areas through the use of "icons" representing each TAA area associated with the LNAV procedure (See Fig. 5-4-9). These icons are depicted in the plan view of the approach chart, generally arranged on the chart in accordance with their position relative to the aircraft's arrival from the en route structure. The WP, to which navigation is appropriate and expected within each specific TAA area, will be named and depicted on the associated TAA icon. Each depicted named WP is the IAF for arrivals from within that area. TAAs may not be used on all LNAV procedures because of airspace congestion or other reasons.

(f) Cold Temperature Limitations. A minimum temperature limitation is published on procedures which authorize Baro-VNAV operation. This temperature represents the airport temperature below which use of the Baro-VNAV is not authorized to the LNAV/VNAV minimums. An example limitation will read: "Baro-VNAV NA below -20°C(4°F)." This information will be found in the upper left hand box of the pilot briefing.

Note: *Temperature limitations do not apply to flying the LNAV/VNAV line of minima using approach certified WAAS receivers.*

(g) **WAAS Channel Number/Approach ID.** The WAAS Channel Number is an equipment optional capability that allows the use of a 5-digit number to select a specific final approach segment. The Approach ID is an airport unique 4-letter combination for verifying selection of the correct final approach segment, e.g. W-35L, where W stands for WAAS and 35L is runway 35 left. The WAAS Channel Number and Approach ID will be displayed in the upper left corner of the approach procedure pilot briefing.

VII. FAR 91.185 IFR Operations: Two-Way Radio Communications Failure

(a) *General.* Unless otherwise authorized by ATC, each pilot who has two-way radio communications failure when operating under IFR shall comply with the rules of this section.

(b) *VFR conditions.* If the failure occurs in VFR conditions, or if VFR conditions are encountered after the failure, each pilot shall continue the flight under VFR and land as soon as practicable.

(c) *IFR conditions.* If the failure occurs in IFR conditions, or if paragraph (b) of this section cannot be complied with, each pilot shall continue the flight according to the following:

(1) *Route.*

(i) By the route assigned in the last ATC clearance received;

(ii) If being radar vectored, by the direct route from the point of radio failure to the fix, route, or airway specified in the vector clearance;

(iii) In the absence of an assigned route, by the route that ATC has advised may be expected in a further clearance; or

(iv) In the absence of an assigned route or a route that ATC has advised may be expected in a further clearance, by the route filed in the flight plan.

(2) *Altitude.* At the highest of the following altitudes or flight levels for the route segment being flown:

(i) The altitude or flight level assigned in the last ATC clearance received;

(ii) The minimum altitude (converted, if appropriate, to minimum flight level as prescribed in § 91.121(c)) for IFR operations; or

(iii) The altitude or flight level ATC has advised may be expected in a further clearance.

(3) *Leave clearance limit.*

(i) When the clearance limit is a fix from which an approach begins, commence descent or descent and approach as close as possible to the expect-further-clearance time if one has been received, or if one has not been received, as close as possible to the estimated time of arrival as calculated from the filed or amended (with ATC) estimated time en route.

(ii) If the clearance limit is not a fix from which an approach begins, leave the clearance limit at the expect-further-clearance time if one has been received, or if none has been received, upon arrival over the clearance limit, and proceed to a fix from which an approach begins and commence descent or descent and approach as close as possible to the estimated time of arrival as calculated from the filed or amended (with ATC) estimated time en route.

INDEX

A

Abbreviations, 119, 144–149
Accuracy of GPS, 17, 18
  with LAAS, 19
  with WAAS, 19
Acronyms, 119, 144–149
ADF, 106, 107
ADS-B (see Automatic dependent surveillance-broadcast)
Aeronautical Information Manual (AIM), 12, 159–209
  GPS excerpts from, 159–181
  instrument approach procedure chart excerpts from, 186–209
  on RAIM capability, 98, 99
  WAAS excerpts from, 181–186
AFD (see Airport/Facilities Directory)
AFM (see Aircraft flight manual)
AIM (see Aeronautical Information Manual)
Air Traffic Control (ATC):
  amended clearances by, 62
  broadcast access by, 124
  electrical failure procedures expected by, 103
  emergency altitudes expected by, 101
  and en route operations, 68
  flight plan changes by, 62
  and refusal of distracting clearances, 67–68
  weather diversions from, 70–71
Airborne weather radar, 121
Aircraft flight manual (AFM), 57–58
Airport/Facilities Directory (AFD), 59
Alaskan Capstone Project, 125
Along track distance (LTD), 149
Alternate airports:
  in DUATS flight planning, 51–52
  with WAAS, 49
Altitude:
  barometric, 151
  decision, 20
  emergency, 101–102
  geometric, 153

Altitude (*Cont.*):
  minimum descent, 20, 155–156
  minimum en route, 102
  minimum safe, 108
  off-route obstacle clearance, 102
  pressure, 156
  vertical navigation features, 31–32
AOPA Pilot, 98–99
Apollo GPS units, 23, 25
  accessing major GPS functions with, 26
  creating/activating flight plans with, 61
  moving map activation with, 67
  WAAS-capability of, 116, 117
Approach charts:
  AIM excerpts on, 186–209
  new items on, 81, 84
  organization of, 64
  overlay, 75–79
  RNAV minima, 77, 82–83
  stand-alone, 77, 80–81
  WAAS, 20
  waypoint symbols on, 84–85
Approaches:
  amendments and missed approach vectors (flight lesson 5), 5–6, 135–136
  anticipating, 47, 50–51
  autopilot advantage in, 88
  changes in, 91–92
  do-it-yourself checklists for, 86, 87
  final course, 93–94
  keystroke briefings for, 88–89
  missed, 5–6, 94, 135–136
  new approach chart items, 81, 84
  overlay, 5–6, 75–79, 133–134
  overlay and stand-alone approaches (flight lesson 4), 5–6, 133–134
  planning of VFR vs. IFR, 68
  precision approaches with WAAS (flight lesson 6), 5–6, 136–137
  RNAV approaches with WAAS, 77, 82–83
  RNAV minima, 85–86
  scaling during, 89–91

Approaches (*Cont.*):
  setting up, 87–88
  stand-alone, 5–6, 77, 80–81, 133–134
  vectored, 92–93
  WAAS, 85–86
  waypoint symbols, 84–85
Area navigation (RNAV), 14, 149–150
Area navigation (RNAV) approach
  configuration, 150–151
ATC (*see* Air Traffic Control)
Automatic dependent surveillance-broadcast
  (ADS-B), 123–125
Autopilot:
  for approaches, 88
  in training aircraft, 8
Autoselection (frequency), 30
Autosequencing (waypoints), 29
Availability, 151
  selective, 18, 142–143
  system, 158

B
Background Briefings, 6–7
  flight checks with GPS (briefing 7–8), 6–7,
    138–139
  GPS approaches (briefing 3–4), 6, 132–133
  GPS background, data entry, and flight
    procedures (briefing 1–2), 6, 129–130
  purpose of, 127
  time allowance for, 127
Backup navaids, 98, 99
Backup satellites, 97
Baro, 14
Baro-aiding, 14, 151
Barometric altitude, 151
Bendix/King GPS units, 23, 24
  accessing major GPS functions with, 26
  creating/activating flight plans with, 61
  expiration screen of, 60
  flight plan function with, 29
  keystroke briefing for, 89
  moving map activation with, 67
  starting screen of, 59
  weather information downlinks with, 122
Black Box Practice, 9, 10
  approaches, 95
  check rides, 114
  emergency procedures, 102
  flight plan, 22, 35, 55, 73
  start-up checks, 73
Blindfold cockpit check, 10

C
CDI needle, 89–90, 93
CHART shortcut, 63
Charts:
  approach (*see* Approach charts)

Charts (*Cont.*):
  for departure procedures, 43
  organization of, 63–64
  viewing, 121
Check rides (*see* Instrument Rating Practical
  Test check rides)
Chelton FlightLogic EFIS, WAAS-capability of,
  116, 118
Clearances:
  changes in, 70
  copying, 62–63
Climb, VNAV rate calculation for, 31
Cockpit checks:
  blindfold, 10
  GPS, 58–62
Cockpit resource management (CRM), 64
Cockpit(s):
  glass, 116, 117
  organization of, 63–64
Communication, GPS interface with, 25
Control station, GPS, 12
Controller-pilot data link communications
  (CPDLC), 151
CRM (cockpit resource management), 64
CSC DUATS, 37, 38
Currency of database, 59

D
DA (decision altitude), 20
Data entry and retrieval, 7–9, 58–60
Database currency, 59
Datalink weather, 121–122
Decision altitude (DA), 20
Defaults, for nearest airport runways, 31
Departure procedures (DPs):
  changing, 65
  in flight planning with DUATS,
    43–46
  instrument, 155
  locating charts for, 43
Departures:
  flight lesson 2, 3–4, 130–131
  flight lesson 3, 3–4, 131–132
  obstacle clearance during, 65, 67
  preparations for, 57–64
  takeoff checks, 64–67
Descent, VNAV rate calculation for, 31,
  72–73
Desired track (DTK), 14, 64, 152
Differential GPS (DGPS), 19, 152
Direct User Access Terminal System (DUATS),
  143–144
  accessing, 38
  alternate choices with, 51–52
  anticipating GPS approach with, 47, 50–51
  backup VOR stations with, 38–39
  departure procedures with, 43–46

Direct User Access Terminal System (*Cont.*):
  filing GPS flight plans with, 53
  GPS planning with, 37–53
  personal minimums with, 52–53
  RAIM predictions from, 100
  rehearsing route worked out by, 43
  sample of GPS flight plan with, 39–42
  STARs with, 47
  weather factors with, 51
Direct-to, checklist for, 28
Distractions:
  future decrease in, 117, 118
  from moving maps, 67
  during takeoff, 65
  with VNAV, 72–73
Diversions, 70–71
DME, 106, 107
Docking stations, 8–9
Do-it-yourself checklists:
  for approaches, 86, 87
  for direct-to, 28
  for flight planning, 28
  for operation with GPS, 26–29
  for power on, 27
DPs (*see* Departure procedures)
DTC DUATS, 37, 38
DTK (*see* Desired track)
DUATS (*see* Direct User Access Terminal System)

E
EGPWS (enhanced ground proximity warning
    system), 122
Electrical failure, complete, 102–103
Electronic glideslope, 91
Emergencies, 97–103
  altitude during, 101–102
  complete electrical failure, 102–103
  GPS outage, 98–100
  and importance of logging times, 101
  two-way radio communications failure,
    101
En route charts, organization of, 63
En route domestic, 152
En route GPS flying, 57, 67–73
  clearance changes, 70
  distractions, 67–68
  diversions, 70–71
  flight lesson 2, 3–4, 130–131
  flight lesson 3, 3–4, 131–132
  holding, 71–72
  messages, 68–69
  moving map, 67
  nearest (NRST) function, 71
  operations, 68, 152
  and VNAV distractions, 72–73
  and VOR—GPS differences, 69
En route oceanic and remote, 152

En route operations, 152
Enhanced ground proximity warning system
    (EGPWS), 122
Entry of GPS data, 7–8, 58–60
Equipment, GPS, 1
  classes of, 153–154
  costs of, 116
  creating/activating flight plans with, 61
  docking stations, 8–9
  handheld/yoke-mounted, 17, 57, 115
  installation requirements for, 16
  legality of, 107
  moving map activation, 67
  in operation (*see* Operation with GPS)
  removable, 8
  (*See also* Technology, GPS)
Estimated time en route (ETE), 101, 152
Estimated time of arrival (ETA), 101, 152
ETA (*see* Estimated time of arrival)
ETE (*see* Estimated time en route)
Examiners, check ride, 105

F
FAA (*see* Federal Aviation Administration)
FAR Part 135 pilots, checklist changes by, 34
FAWP (final approach waypoint), 94
Federal Aviation Administration (FAA):
  and Alaskan Capstone Project, 125
  and check ride examiners, 105
  commitment to GPS improvements, 18
  GPS AIM excerpts, 159–209
  instrument approach procedure charts AIM
    excerpts, 186–209
  selective availability statement by, 142–143
  VOR backup stations required by, 38–39
  WAAS AIM excerpts, 181–186
Filing GPS flight plans, 53
Final approach, 93–94
Final approach waypoint (FAWP), 94
Flat map navigation, great circle navigation vs.,
    14, 15
Flight lessons, 3–6, 127–140
  Background Briefings for, 6–7
    (*See also* Background Briefings)
  basic GPS functions (lesson 1), 127–129
  departures and en route GPS—Part I (lesson 2),
    130–131
  departures and en route GPS—Part II (lesson 3),
    131–132
  GPS approaches—Part II: Approach
    amendments and missed approach vectors
    (lesson 5), 135–136
  GPS approaches—Part III: Precision approaches
    with WAAS (lesson 6), 136–137
  GPS instrument approaches—Part I: Overlay
    and stand-alone approaches (lesson 4),
    133–134

Flight lessons (*Cont.*):
  long IFR cross-country with GPS (lesson 7), 137–138
  pre- and postflight briefings for, 6
  preparation for GPS-based FAA instrument check ride (lesson 8), 139–140
  time allowance for, 127
Flight management system, 152–153
Flight plan function, 61
Flight planning, 37–55
  alternates in, 51–52
  anticipating GPS approach in, 47, 50–51
  check ride review of, 107
  checklist for, 28
  departure procedures, 43–46
  with DUATS, 37–53
  filing GPS flight plans, 53
  personal minimums in, 52–53
  rehearsal of route, 43
  sample of, 39–42
  standard terminal arrival procedures, 47–49
  VOR backup in, 38–39
  weather factors in, 51
  without DUATS, 53–55
Flight plans:
  advance preparation of, 5
  based on Victor airways, 69
  docking station practice flying of, 8
  entering, 28, 29
  sample of, 39–42
  storing, in GPS unit, 61, 62
Flight technical error (FTE), 153
"Fly it first" mentality, 33
Fly-by waypoint, 85, 153
Flying ability test, 111–112
Fly-over waypoint, 85, 153
Frequencies:
  autoselection of, 30
  tracking, 64
FTE (flight technical error), 153
Future of GPS, 115–125
  ADS-B, 123–125
  airborne weather radar, 121
  chart viewing, 121
  costs of equipment, 116
  datalink weather, 121–122
  display distractions, 117, 118
  glass cockpits, 116, 117
  lightning detection, 122
  multifunction displays, 120–121
  PFD design, 118–120
  terrain awareness, 122
  traffic advisories, 122–123
  transitioning to WAAS, 115–118

G
Garmin GPS units, 23, 24
  accessing major GPS functions with, 26

Garmin GPS units (*Cont.*):
  creating/activating flight plans with, 61
  moving map activation with, 67
  weather information downlinks with, 121
Geometric altitude, 153
Glass cockpits, 116, 117
Global navigation system landing system (GLS), 153
Global positioning system (GPS), 1
  accessing major functions of, 26
  accuracy of, 17–19
  availability of, 18, 142–143
  basic GPS functions (flight lesson 1), 3–4, 127–129
  equipment classes A, B, and C., 153–154
  interfaces with, 25
  monitoring stations, 12
  nonprecision approaches with, 20
  outage of, 98–100
  receivers, 12–14
  satellites, 12, 13
  terms used with, 14
    (See also Acronyms; Glossary)
  transition to, 12
  typical check ride questions about, 108–111
  using manuals for, 11–12
  VFR use of, 15, 16
  and WAAS, 18–19
  and WAAS-approach capability, 20–22
    (See also specific topics)
Glossary, 149–159
GLS (global navigation system landing system), 153
GNS 430, 530, 9
"Gotcha" button, 33
GPS (see Global positioning system)
GPS Trainer 2.0, 9
GPS/NAV interfaces, 25
GPS/NAV selector switch, 33
Graves, Alexis, 28
Great circle navigation, 14, 15, 38
Ground stations, WAAS, 18
Ground uplink stations (GUSs), 18

H
Handheld GPS units, 115
  for cross-country flights, 57
  lack of RAIM with, 17
HDOP (horizontal dilution of precision), 154
Holding, 71–72
Horizontal dilution of precision (HDOP), 154

I
IAF (initial approach fix), 154
IAP (see Instrument approach procedure)
IF (intermediate fix), 155
IF/IAWP (intermediate fix/initial approach waypoint), 154

IFR equipment, certifications for, 16
IFR flying:
 accuracy of GPS for, 18
 cross-country, 37, 137–138
 database for, 17
 GPS as supplemental system for, 38–39
 GPS unit use in, 16, 17
 long IFR cross-country with GPS (lesson 7),
  137–138
 minimum altitude for, 101–102
 personal minimums for, 52–53
 planning, without DUATS, 53–55
 rating for, 105, 106
 and yoke-mounted/handheld GPS units, 57
IFWP (intermediate fix waypoint), 154
IGEB (Interagency GPS Executive Board), 18
IMC (instrument meteorological conditions), 155
Initial approach fix (IAF), 154
Instructor notes:
 GPS self-tests/preflight checks, 60–61
 for lessons 1-3, 3–4
 for lessons 4-6, 7
 nearest airport defaults, 31
Instrument approach procedure (IAP), 154–155
 setting up, 87–88
 and wind direction, 47
Instrument departure procedures, 155
Instrument meteorological conditions (IMC), 155
Instrument Rating Practical Test check rides,
  105–114
 common GPS deficiencies in, 113–114
 determining minimum safe altitudes, 108
 examiners for, 105
 flight planning review, 107
 flying ability test, 111–112
 with GPS-equipped aircraft, 3, 106–107
 and legality of GPS equipment, 107
 nonprecision approaches in, 106
 oral examination, 108–111
 preparation for (flight lesson 8), 139–140
 typical GPS oral questions, 108–111
Instrument Rating Practical Test Standards, 102
Integrity, 155
Interagency GPS Executive Board (IGEB), 18
Interfaces with GPS, 25
Intermediate fix (IF), 155
Intermediate fix waypoint (IFWP), 154
Intermediate fix/initial approach waypoint
  (IF/IAWP), 154
Internet aviation weather services, 53

J
Jamming, 155
Jex, Robert, 98–99

K
K (identifier letter), 14
Keystroke briefings, 88–89

Keystrokes, learning, 7–9
 for approaches, 88–89
 black box practice for, 9, 10
 practice in, 89

L
LAAS (*see* Local Area Augmentation System)
Lateral navigation (LNAV), 86, 155
Lateral navigation/vertical navigation
  (LNAV/VNAV) approach, 86
Latitude, 59, 60
Legality of GPS, 107
Lightning detection, 122
LNAV (*see* Lateral navigation)
LNAV/VNAV approach, 86
Local Area Augmentation System (LAAS), 19
 FAA commitment to, 18
 precision approaches promised by, 20
 transitioning to, 115
Logging of times, 101
Longitude, 59, 60
LPV DA approach, 85–86
LTD (along track distance), 149

M
M (*see* Message annunciator)
Manuals:
 aircraft flight manual, 57–58
 GPS, use of, 11–12
Maps:
 flat, 14, 15
 moving, 25, 67
Mask angle, 155
Mastering Instrument Flying (Henry Sollman), 52,
  55, 112
MAWP (missed approach waypoint), 94
MDA (*see* Minimum descent altitude)
MEA (minimum en route altitude), 102
Message annunciator (M or MSG), 16, 68–69,
  98
Messages, 68–69
 RAIM, 16, 68, 98–100
 VNAV, 31
 WAAS, 100
MFDs (*see* Multifunction displays)
Military GPS level, 18
Minimum descent altitude (MDA), 20,
  155–156
Minimum en route altitude (MEA), 102
Minimum safe altitudes (MSAs), 108
Minimums:
 personal vs. published, 52–53
 RNAV minima, 85–86
Missed approach point, 156
Missed approach waypoint (MAWP), 94
Missed approaches, 94, 156
Mistakes, correcting, 32–35
Monitoring stations, GPS, 12

Moving maps:
distraction by, 67
for en route flying, 67
GPS interface with, 25
MSAs (minimum safe altitudes), 108
MSG (*see* Message annunciator)
Multifunction displays (MFDs), 120–123

N
Navigation:
GPS interface with, 25
great circle vs. flat map, 14, 15
NDB approaches, 106, 107
Nearest (NRST) function, 30, 31, 71
Nonprecision approach operations, 156
Nonprecision approaches:
in check rides, 106
MDA indicating, 20
types of, 20
NRST function (*see* Nearest function)

O
OBS mode, 28, 93
Off-route obstacle clearance altitude (OROCA),
102
Operation with GPS, 23–35
correcting mistakes in, 8–9, 32–35
direct-to mode, 27–28
do-it-yourself checklists for, 26–29
en route, 68
entering flight plans, 28, 29
frequency autoselection, 30
nearest (NRST) function, 30, 31
OBS mode, 28
powering up, 25–27
vertical navigation, 31–32
waypoint alerting and turn anticipation, 29, 30
waypoint autosequencing, 29
Oral examination (check rides), 108–111
Organization of cockpit, 63–64
OROCA (off-route obstacle clearance altitude),
102
Outage, GPS, 98–100
Overlay approaches, 5–6, 20, 75–79, 133–134

P
Panel-mounted GPS units, 57, 115
Personal minimums, 52–53
PFD (*see* Primary flight display)
Portable transceivers, 103
Position data, GPS, 13–14
Postflight briefings, 6
Power on, 25–27
Power outages, 97
PPS (precise positioning service), 18
Practical test (*see* Instrument Rating Practical
Test check rides)

Practical Test Standards (PTSs), 106, 110
Precise positioning service (PPS), 18
Precision approaches:
and accuracy of GPS, 18
with LAAS, 20
WAAS capability for, 20–22
WAAS charts, 20
WAAS (flight lesson 6), 5–6, 136–137
Predictions, RAIM, 100
Preflight briefings, 6
Preflight procedures, 57–64
before-takeoff checks, 64
cockpit organization, 63–64
copying clearances, 62–63
GPS cockpit checks, 58–62
walk-around inspection, 57–58
Pressure altitude, 156
Primary flight display (PFD), 118–120
Priorities in flying, 33
Pseudo-range, 156
PTSs (*see Practical Test Standards*)

R
RAIM (*see* Receiver autonomous integrity
monitoring)
Ratings, pilot, 105, 106
Receiver autonomous integrity monitoring
(RAIM), 16–17, 97–98
defined, 156–157
predictions from, 100
warnings from, 98
Receivers:
GPS, 12–14
RAIM function in, 16–17, 97–98
VFR-only, 14, 15
WAAS-certified, 19
Rehearsing routes, 43
Required navigation performance, 157
Resources for information/supplies/equipment,
9, 141–142
Retrieval of GPS data, 7–8
RMIs, 106–107
RNAV (*see* Area navigation)
RNAV approach configuration (*see* Area
navigation approach configuration)
RNAV approaches with WAAS, 77,
82–83
RNAV minima, 85–86

S
Safety pilot, 10
Satellites:
backup, 97
GPS, 12, 13, 97
WAAS, 18
Scaling (approaches), 89–91
Sectional charts, organization of, 64

Selective availability, 18, 142–143
Setting up approaches, 87–88
Shortcuts:
    blindfold cockpit check, 10
    CHART, 63
    for check rides, 110
    practicing VNAV, 32
    VFR-only GPS as cross-check, 17
Simulated flying:
    with docking station at home, 8
    in parked aircraft, 9
Sollman, Henry, 52, 55, 112
Sources of information/supplies/equipment,
        141–142
Stand-alone approaches, 5–6, 20, 77, 80–81,
        133–134
Stand-alone GPS navigation system, 157
Standard terminal arrival (STAR), 34, 47,
        157
Student note (nearest airport defaults), 31
Supplemental air navigation system, 157
Syllabus:
    Background Briefing 1-2, 129–130
    Background Briefing 3-4, 132–133
    Background Briefing 7-8, 138–139
    Flight lesson 1, 127–129
    Flight lesson 2, 130–131
    Flight lesson 3, 131–132
    Flight lesson 4, 133–134
    Flight lesson 5, 135–136
    Flight lesson 6, 136–137
    Flight lesson 7, 137–138
    Flight lesson 8, 139–140
Symbols, waypoint, 84–85
System availability, 158

T
Takeoff checks, 64–67
Takeoff time, logging, 101
TASs (*see* Traffic advisory systems)
TAWS (terrain awareness and warning system),
        122
TCAS (traffic alert and collision avoidance
        system), 123, 158
Technology, GPS, 11–22
    accuracy of, 17–18
    common abbreviations/acronyms for, 119,
        144–149
    and great circle navigation, 14, 15
    LAAS, 19
    manuals for, 11–12
    monitoring stations, 12
    RAIM function in, 16–17
    receivers, 12–14
    satellites, 12, 13
    WAAS, 18–22
Terminal area operations, 158

Terrain awareness and warning system (TAWS),
        122
Time data, GPS, 13
Times, logging, 101
TIS (*see* Traffic information system)
TK (*see* Track)
Track angle error, 149, 158
Track (TK), 14, 64, 152, 158
Traffic advisory systems (TASs), 122–123
Traffic alert and collision avoidance system
        (TCAS), 123, 158
Traffic information system (TIS), 123, 124
Transceivers, portable, 103
Transition, 158
Turn anticipation, 30
Two-way radio communications failure:
    altitude for completing flight with,
        101–102
    and familiarity with DPs, 43
    FAR 91.185 IFR operations, 209–210
    handling, 101

U
Universal access transceiver (UAT), 124
User-defined waypoints, 14, 158

V
Vectored approaches, 92–93
Vertical navigation (VNAV), 31–32, 158
    distractions with, 72–73
    practicing, 32
VFR flying:
    duplicating GPS and VOR usage for, 34
    GPS use for, 15, 16
    when aborting IFR flights, 52–53
Victor airways, 69
Visual descent point, 159
Visual meteorological conditions (VMC), 10,
        159
VNAV (*see* Vertical navigation)
VOR navigation:
    as backup for GPS flights, 38
    differences between GPS and, 69
    as STAR basis, 34
    when unclear messages appear, 69

W
WAAS (*see* Wide Area Augmentation System)
WAAS ground reference stations (WRSs), 18
Walk-around inspection, 57–58
Waypoint(s), 159
    alerting to, 29, 30
    autosequencing of, 29
    defined, 14
    on final approach, 93
    fly-by, 85, 153
    fly-over, 85, 153

Waypoint(s) (*Cont.*):
  intermediate fix, 154
  intermediate fix/initial approach, 154
  symbols for, 84–85
  user-defined, 14, 158
  for vectored approaches, 92–93
Weather:
  airborne weather radar, 121
  datalink weather, 121–122
  diversions for, 70–71
  DUATS briefings on, 47, 51
  in flight planning, 51
  Internet aviation services, 53
Wide Area Augmentation System (WAAS),
    18–19, 159
  approach capabilities with, 20–22
  coverage of, 19

Wide Area Augmentation System (WAAS) (*Cont.*):
  as differential GPS, 19
  electronic glideslope with, 91
  FAA commitment to, 18
  and GPS accuracy, 19
  GPS alternates with, 49
  ground stations for, 18
  and RAIM function, 100
  satellite constellation for, 18
  transitioning to, 115–118
Wide area master stations (WMSs), 18
WRSs (WAAS ground reference stations), 18

Y
Yoke-mounted GPS units:
  for cross-country flights, 57
  lack of RAIM with, 17

ABOUT THE AUTHORS

Phil Dixon holds ATP, Flight Engineer, and Flight Instructor Certificates with Instrument and Multi-Engine ratings. He has served as an FAA Designated Pilot Examiner and as an Aviation Safety Counselor. Mr. Dixon graduated from the University of Tennessee (Knoxville) in 1982 with a Bachelor of Science degree in communications. He is a member of AOPA and lives in the St. Louis area.

Sherwood Harris is coauthor (with Henry Sollman) of the well-received McGraw-Hill book, *Mastering Instrument Flying*, now in its third edition. A former Navy carrier pilot, he holds an ATP and Flight Instructor Certificates with Single, Multi-Engine, and Instrument ratings. Mr. Harris was a part-time flight instructor at Danbury, Connecticut, and White Plains, New York, for many years. A member of AOPA, he is a retired Reader's Digest senior staff editor and makes his home in Oxford, Mississippi.